Glass in Northwest Ohio

On the cover: The master glassblower and gatherer team for over 3,000 years dominated the glass industry. A young northwest Ohio glassblower, Michael Owens, started a revolution in the 1890s. Here we see the glassblower and gatherer working with Owens's first evolutionary step, the automated mold. Full automation would result in the elimination of the glassblower and gatherer. Automation, however, reduced prices and caused an employment boom in the glass industry. (Courtesy of the Library of Congress.)

GLASS IN NORTHWEST OHIO

Quentin R. Skrabec Jr., Ph.D.

ARCADIA
PUBLISHING

Copyright © 2007 by Quentin R. Skrabec Jr., Ph.D.
ISBN 978-0-7385-5111-1

Published by Arcadia Publishing
Charleston, South Carolina

Printed in the United States of America

Library of Congress Catalog Card Number: 2006939091

For all general information contact Arcadia Publishing at:
Telephone 843-853-2070
Fax 843-853-0044
E-mail sales@arcadiapublishing.com
For customer service and orders:
Toll-Free 1-888-313-2665

Visit us on the Internet at www.arcadiapublishing.com

To Our Lady of Toledo and my grandmother Finnerty,
orphan and glassworker.

CONTENTS

ACKNOWLEDGMENTS

No such book can be created without the help of librarians, local historians, and archivists. I was blessed with an exceptional group of these in the northwest Ohio area. Clearly, one of the most professional and helpful of these groups was the staff of the Ward M. Canaday for Special Collections at the University of Toledo. I would like to highlight Kimberly Browniee, who has collaborated with me on a number of projects. Similarly, I wish to thank the staff of the local history department of the Toledo-Lucas County Library. Michael Lora, James Marshall, and Greg Miller, in particular, were extremely knowledgeable and helpful. Special thanks also go to Nicole Rivette and Julie McMaster of the Toledo Museum of Art, which is the mecca of glass research. Many of the local librarians were exceptional in their service, such as Anne Bushel and Charles Butcher of the Rossford Library. In addition, I worked with local glass historian Lawrence Bridenbaugh of Rossford. Jeffrey Winkle and Penelope Justice of the Fostoria Kaubisch Memorial Public Library were extremely helpful and patient. Special thanks also go to Lee McLaird, curator of rare books and special collections at Bowling Green State University. Another integral aid was Paulette Weiser, curator of the Hancock Historical Society, which supplied several Findlay pictures from its outstanding archives and museum. My graduate research assistant, Gongala V. S. Sirkanth, was instrumental in the research of this book. Finally, thanks to Dr. Freed and Dr. Sears, who continue to encourage the research of local industries at the University of Findlay.

INTRODUCTION

In the 1850s, railroads and canals made northwest Ohio a key lake port and distribution center. By 1870, northwest Ohio had risen out of the black swamp as an agricultural center, but the 1880s were frustrating, as the area lost in its effort to become the major distribution center for iron ore to Cleveland and grain to Chicago. Toledo did have seven major railroads though, which only Chicago could rival, making things like coal as cheap as they were in Pittsburgh and ultimately playing a key role in bringing glass to northwest Ohio.

New England was the center of glassmaking from the 1700s to the 1850s, but the great hardwood forest that fueled it was fast disappearing. The New England Glass Company became the cradle of American glassmakers, with a stream of glassmakers evolving that included William Libbey, who took control of the company in 1852. Deforestation caused a major industry shift in fuel and geography by 1850. The coal of Pittsburgh and West Virginia offered a cheap fuel in the 1860s and drew many glass companies from New England. Natural gas discoveries in Pittsburgh and West Virginia offered a new fuel in the 1870s. Ultimately gas was cheaper than coal and supplied a more even heat. It also was much cleaner, which favored the tableware and art glass industry.

Traditionally, the industry strived where fuel was cheap. By 1880, Pittsburgh was the world's glass center with 44 glasshouses. There were no glasshouses in northwest Ohio, but it was connected to the great glass centers of New England and Pittsburgh by the firm of Card and Hubbard, which had been shipping high-quality sand from Monclova to them since before the Civil War. Then, in the early 1880s, natural gas was discovered in Findlay, Ohio. Over 100 glass companies came to two gas fields, one field two by five miles around Findlay and another one by two miles centered in North Baltimore, Ohio. At first the gas deposit seemed endless, inspiring towns like Findlay, Tiffin, North Baltimore, Bowling Green, and Fostoria to offer free gas to glass companies. Toledo piped the gas in. The glass rush was facilitated by a highly mobile glass industry. Almost all of the glasshouses could be packed up and moved by train to a new location, usually requiring 25 to 50 carloads. Most were wooden structures, making them very mobile, but they commonly burnt down. The northwest Ohio gas supplies lasted less than five years, but it changed the region forever.

In 1886, Findlay's population was 5,000; a year later, Findlay's population was 15,000. Railroad crossings such as North Baltimore became towns over night. Northwest Ohio was a railroad center, and the monopolistic price controls of the railroads actually created higher freight rates for the big cities of New York, Pittsburgh, Boston, and Baltimore. Toledo and Bowling Green freight cost 10¢ per 100 pounds to the western growth markets versus 14¢ to 17¢ for the bigger eastern cities. This freight advantage was particularly important to the window-glass companies, who rushed to northwest Ohio. Low-cost freight also made alternative fuels such as coal as cheap in Toledo as in Pittsburgh.

Within a few years, almost 100 glass companies had rushed into northwest Ohio. By 1890, Findlay had 25 companies, Tiffin had 6, Fostoria had 8, and Bowling Green had 8. By 1892, the gas pressure had dropped significantly in northwest Ohio, so again many followed the cheap fuel sources. About 50 companies packed up and left for Indiana and better gas supplies. Most of the remaining northwest Ohio glasshouses would close within 15 years, as the natural gas ran out, but several would stay and revolutionize the glass industry forever. Edward Libbey would bring the grandfather of the American glass industry—New England Glass—to northwest Ohio. His decision was based on the transportation roots of the area rather than the gas. Libbey designed his Toledo plant to function off gas, oil, or coal. Libbey would also bring a young genius, Michael Owens, from West Virginia to lead this revolution.

Glassmaking prior to Michael Owens and Edward Libbey had remained unchanged for 3,000 years. Known as the Venetian method, it consisted of hand blowing molten glass at the end of a four-foot-long iron pipe known as a blowpipe. Molten glass was melted from sand, soda, recycled glass, and other ingredients in a pot. A wood-fired furnace normally held 10 pots of molten glass. A crew known as a gang or shop worked each pot at each of 10 furnace openings. A typical northwest Ohio glasshouse would have 10 pot furnaces in the blown-glass industry. A gatherer would gather a gob of molten glass from the pot for the blower. The glassblower was the master and was in charge of the gang. The gatherer or assistant blower would use another metal rod called a pontil to remove the piece for further finishing or reheating. Several boys were on a crew to open and close molds if used, to take finished pieces to the annealing furnace known as a lehr, to feed the furnace fires, and to prepare tools. Commonly there would be five to seven workers in a gang. The nature of the glass industry in the 1880s was very different than today. There were several distinct segments of the glassmaking industry. The oldest segment was the green bottle industry. This segment produced handblown bottles, lamps, and tableware. The word green related to the color of cheap glass made green by impurities. The largest segment was the flint-glass branch that produced pressed and blown tableware as well as other handblown items. Flint glass was clear and heavy as the result of large amounts of lead oxide, but the segment also included the cheaper lime glass, lacking lead for low-cost pressed ware. Lime glass was more readily pressed and had been developed by New England Glass. Flint glass producers also produced expensive cut-glass products for the home. Another segment was the pane or window glass segment, which was flat glass produced from handblown cylinders. Window glass quality was poor due to the distortion of flattening. The polished plate glass segment used cast-glass slabs that were highly polished and could be found in wealthier homes. The unions formed according to the industrial segment, such as the Flint Glass Workers, American Window Glass Workers, and the Green Bottle Blowers.

The vast majority of laborers were boys between the ages of 10 and 15. Ohio law barred children less than 12 years of age, but the economic growth of the glass industry required most of these laws to be ignored. The pictures in this book graphically reflect this abuse. These glasshouse children found no support from the union, which ignored them. The lure of jobs for young boys in the glasshouses brought many families to northwest Ohio. Orphans were brought in from the big eastern cities to fill the jobs. The shortage of child labor in Toledo caused an influx of women into the industry as well.

Labor trouble at the New York Corning Glass plant required Edison Electric to search for a new supplier of bulbs in 1890. The opportunity came to Libbey Glass through Sol Richardson, who had worked with Edison early on. Libbey rented a plant in Findlay to manufacture handblown bulbs, putting Richardson and Michael Owens in charge. Owens, who had worked as a mold boy, started the industry toward automation in 1892 at the Findlay bulb operation. The Libbey Glass plant in Findlay faced a strike by mold boys at start up, forcing Owens to design an automated mold for light bulb manufacture using a foot pedal. The Libbey Findlay plant had many times the productivity of the Corning Glass bulb plant, but union agreements required the business to go back to Corning when the labor dispute was settled; however, Libbey retained a minor position in the light bulb business. The business took root in northwest Ohio with a bulb assembly plant, Fostoria Incandescent Lamp Company, which was initially supplied by Libbey. Several bulb-blowing factories also came to Fostoria to support the assembly plant.

Owens followed his Findlay success with a marketing victory for Libbey Glass at the Chicago World's Fair in 1893. Libbey Glass had become a premier producer of cut crystal glass. Libbey was a major exhibitor at the fair, against the advice of his whole board of directors, finding his only support in local banker George Ketcham and his young manager Michael Owens. The fair success saved a financially struggling Libbey Glass and built its international reputation.

Libbey formed a new company in 1895, Toledo Glass, for Owens to experiment on automation. Libbey gave the new company a small wooden shed on Ash Street. Libbey brought two other key men to the new company, W. S. Walbridge, his brother-in-law, and engineer William (Emil) Bock. In 1898, Toledo Glass had its first breakthrough with the semiautomatic oil lamp chimney machine, which brought chimney glass producers such as MacBeth-Evans to northwest Ohio. Using a bicycle pump device in 1900, Owens produced a four-ounce petroleum jelly jar using the new device. By 1903, Toledo glass rolled out a fully commercial machine capable of producing 12 beer bottles per minute. The success of the Owens bottle machine led to the formation of Owens Bottle Machine Company. By 1912, the Owens machine could make 50 bottles a minute, changing the industry forever. Owens machines were being used all over the world by 1915, making Toledo the glass capital of the world.

It is little remembered how the Owens bottle machine changed American lives. Prior to the automated glass machine, things like baby bottles, glass jars, soda bottles, milk bottles, and beer bottles did not exist. The immediate fear of the automated machine was the loss of jobs. American glass had been the darling of the McKinley tariffs of the second half of the 1800s, now a machine might undo the job protection of decades. Thanks to the vision of Owens and the union officials, few jobs were lost in the early years, and ultimately a glass boom resulted in the use of cheap bottles and containers. Not only did glass jobs triple form 1880 to 1920, but so did the population of northwest Ohio. The Owens machine would change lives in another ancillary way. In 1913, Owens received a letter from the National Child Labor Committee, which congratulated him on eliminating more child labor than they had through legislation. In 1880, it was estimated that 24 percent of the bottle factories used child labor. By 1919, thanks to the Owens machine that amount was 1.8 percent.

Owens Bottle Company formed to lease machines to bottlers and packagers. By leasing machines, Toledo Glass Company could determine who manufactured what. It allowed Owens Bottle to control the more profitable beer bottle market. The glass market was seen as controlled by the "Toledo Glass Faction," consisting of Edward Libbey, Michael Owens, Clarence Brown, William Walbridge, William Boshart, and John Biggers. This group allowed for interlocking directorships between a large number of glass companies, including Libbey Glass, Toledo Glass, Westlake Machine, and Owens Bottle Company. In addition, Libbey and Owens formed leasing companies throughout Europe and Asia. By 1925, almost every manufactured bottle in the world was produced on an Owens machine, with royalties going to Toledo Glass. The 1929 merger of Owens Bottle and Illinois Glass formed the Owens-Illinois Company, the world's largest container manufacturer. The Toledo Glass Faction literally controlled the world market for bottles, bulbs, and jars. They soon would use their enormous assets to conquer the flat glass industry.

But before conquering the flat glass market, Owens automatic machines would break into two other huge markets—glass light bulbs and glass tubing. Glass production of bottles, bulbs, and tumblers actually increased with the Owens automation, while cylinder window glass manufacturers such as Findlay Window Glass Company, Canastota Glass of Bowling Green, Mambourg Glass of Fostoria, Tiffin Glass, and Toledo Window Glass decreased. Flat glass production was the most labor sensitive of the glass industry segments. Window glass was produced by blowing a large cylinder. The ends of the cylinder are cut off and the cylinder cut lengthwise. Then the glass is heated and flattened to make a flat. It required a great deal of secondary heating and was very fuel driven. Thicker flat glass was known as plate glass and was cast on an iron table, flattened, and then ground to size. Plate glass was used for store windows versus the smaller sections of cylinder glass used for houses.

Edward Ford would break from Pittsburgh Plate Glass, which dominated the plate business, to come to northwest Ohio. Ford's decision to move to northwest Ohio had nothing to do with natural gas, which was gone by 1899, or Owens's automatic machines, which did not produce plate glass. Ford based his decision on the white sand of Monclova and Holland, the abundance of Toledo limestone, and cheap coal via the railroads. One other local resource played a major role—soda ash. Soda ash was important to high-quality plate and window glass and was being purchased from Europe but had recently been found in Wyandotte, Michigan. From Michigan it could easily and cheaply be shipped by railroad to northwest Ohio. In 1930, Libbey-Owens Sheet would merge with Ford to form Libbey-Owens-Ford.

The path to that merger would be based on other revolutionary inventions of Owens and Irving Colburn. Irving Colburn was a poor inventor who dreamed of a simple machine-cast glass process to make window and plate glass. He had struggled for money, technical help, and encouragement until he met Owens in 1910. Owens knew how to mesh dreams with innovation, and he would team up with Colburn to revolutionize the flat plate and window glass market with an automated flat glass machine.

While production faded in the 1920s, glass technology was centered in northwest Ohio. By 1930, 85 percent of the world's glass products were made on Owens bottle machines and Colburn flat glass machines. Remnants of the trust even remain today, with Owens-Illinois making over 50 percent of the world's containers, and it is the world's leading manufacturer of glass containers. While many towns of northwest Ohio no longer produce glass, they remain treasure groves for glass collectors.

One

INDUSTRY ROOTS

The history of most of America's glass industry can be traced in some way to the New England Glass Company, but in a very special way it is the grandfather of northwest Ohio glass. The founder of New England Glass, Deming Jarves, built the American glass industry. Europe had jealously guarded the secrets of high-quality glassmaking, preventing new entry into the industry. The glassmaking nations of Europe had enacted laws forbidding the emigration of glass artisans and glassblowers. The famous Venetian glassmakers for centuries were isolated on a guarded island. In addition, the Venetian government imposed the death penalty on those trying to emigrate.

Jarves in the 1840s smuggled out a number of glass artisans that broke the code of flint glass making. Flint glass was a high-quality cut glass containing 20–28 percent lead, which gave it its weighty feel. Jarves was able to unlock the European monopoly on the recipes for flint glass. One of those smuggled glass experts, Thomas Leighton would later revolutionize the glass industry with American pressed soda glass. In the 1870s, William Libbey, who had worked mainly in sales, took over the struggling New England Glass Company. The company was suffering from soaring fuel costs and competition from pressed glass made by Hobbs-Brochunier of West Virginia. In addition, the glassworkers were pushing for wage increases equal to those of more successful companies of West Virginia.

New England Glass pioneered beautiful colored art glass. The secret to ruby red glass had been lost with the death of Prussian alchemist Johann Kunckel Von Lowenstern in 1680. A New England glassblower accidentally rediscovered it in 1882 when a gold ring fell into a bath of glass creating rich ruby red. Libbey and his glassmakers went on to manufacture an array of red, pink, and peach colors going by such names as Amberina, Agate, and Pomona. William Libbey's son, Edward, would successfully sell Tiffany's of New York in promoting the new glass products. The success of these colored glass products helped New England Glass maintain a few more years with the high cost of New England fuel.

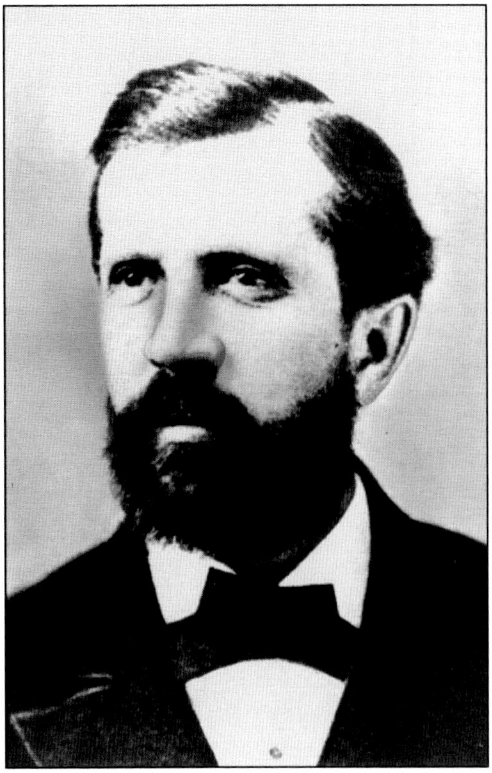

William Libbey was born in Portsmouth, New Hampshire, and could trace his first ancestor to the 1620 landing of the *Mayflower*. In 1855, Libbey started as an importing clerk for the New England firm of Jarves and Commerais, a precursor of New England Glass. Libbey worked his way through various managerial positions until he took ownership in 1867. He may well deserve the title of "father of American glass."

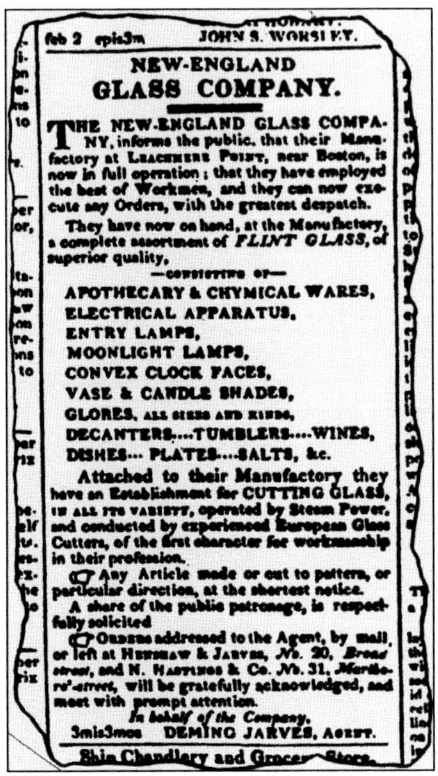

feb 2 epis3m JOHN S. WORSLEY.

NEW-ENGLAND
GLASS COMPANY.

THE NEW-ENGLAND GLASS COMPANY, informs the public, that their Manufactory at LEACHMERE POINT, near Boston, is now in full operation; that they have employed the best of Workmen, and they can now execute any Orders, with the greatest despatch.

They have now on hand, at the Manufactory, a complete assortment of *FLINT GLASS*, of superior quality,

—CONSISTING OF—

APOTHECARY & CHYMICAL WARES,
ELECTRICAL APPARATUS,
ENTRY LAMPS,
MOONLIGHT LAMPS,
CONVEX CLOCK FACES,
VASE & CANDLE SHADES,
GLOBES, ALL SIZES AND KINDS,
DECANTERS....TUMBLERS....WINES,
DISHES... PLATES....SALTS, &c.

Attached to their Manufactory they have an Establishment for CUTTING GLASS, IN ALL ITS VARIETY, operated by Steam Power, and conducted by experienced European Glass Cutters, of the first character for workmanship in their profession.

☞ Any Article made or cut to pattern, or particular direction, at the shortest notice.

A share of the public patronage, is respectfully solicited

☞ ORDERS addressed to the Agent, by mail, or left at HENSHAW & JARVES, *No.* 20, *Broad street*, and N. HASTINGS & Co. *No.* 31, *Marlboro'-street*, will be gratefully acknowledged, and meet with prompt attention.

In behalf of the Company,
3mis3mos DEMING JARVES, AGENT.

Ship Chandlery and Grocery Store.

In 1818, the New England Glass Company took over the assets of the failed Boston Porcelain and Glass Company. The major investor was Deming Jarves, who wanted to move into the fine cut-glass market. Needing flint sand, potash, abundant fuels, and lead, the only lacked component was lead oxide (known as red lead) for the production of flint glass. (Courtesy of the University of Toledo's Ward Canaday Center.)

Jarves was a self-educated glassmaker. Jarves studied European books to develop a practice to manufacture red lead. He succeeded, and New England had a monopoly on red lead for decades. Under Jarves's leadership, New England Glass built a reputation for the finest flint glass. In 1826, Jarves gave up control of the company to Boston and Sandwich Glass Company.

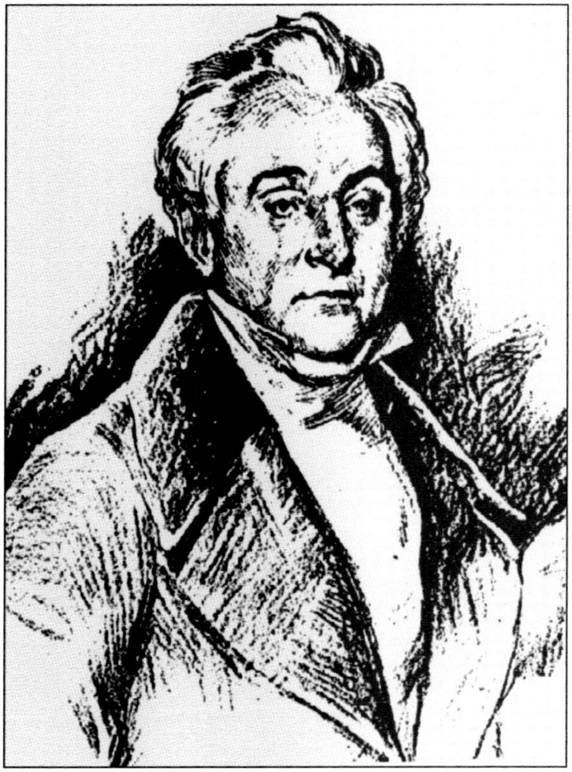

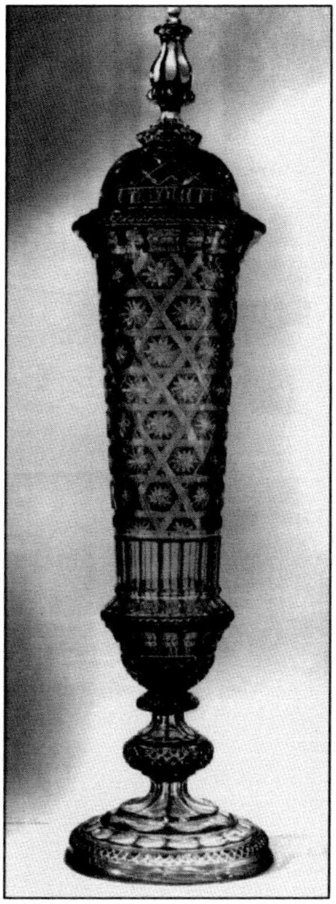

This type of apothecary jar was a popular lead-glass piece of New England Glass. These pieces were hand blown and hand cut with layered ruby glass. Ruby-colored glass was reintroduced to the world after the recipe was lost for centuries.

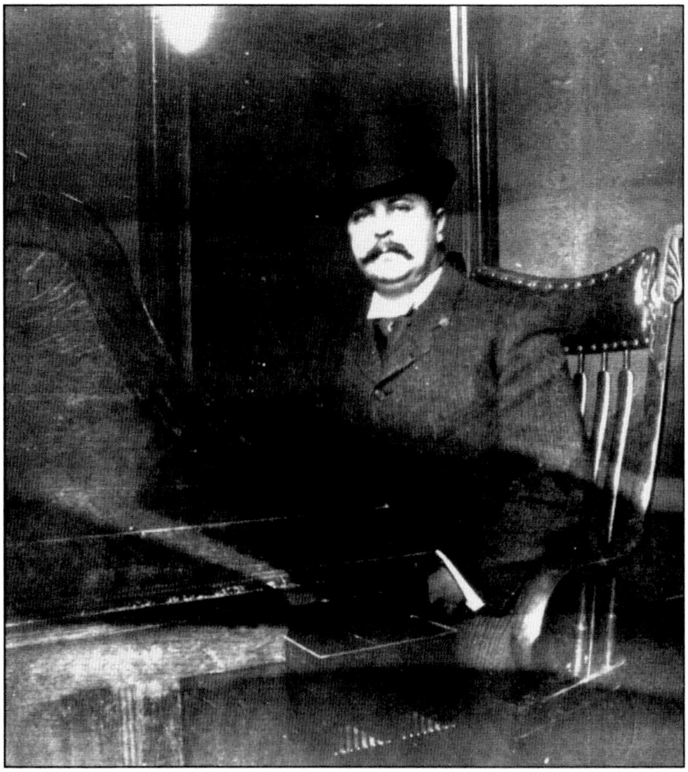

Edward Libbey had joined his father's firm in 1874 at age 19. He showed early successes selling colored cut glass to major outlets. In 1880, he was made a partner as the legal name of the firm became William L. Libbey and Son, but the company still worked as New England Glass. He took over the company after his father's death in 1882.

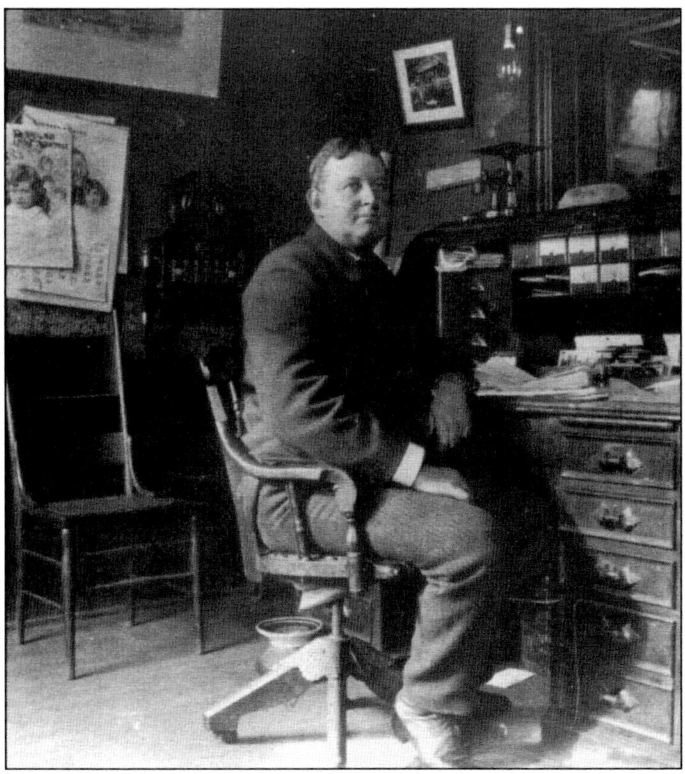

While Edward Libbey worked hard in improving sales at New England Glass, he depended on his young product development manager, Sol Richardson, to improve the struggling company's manufacturing. In the 1880s, Richardson developed a working relationship with Thomas Edison at Menlo Park. Richardson did some of the earliest production work on the electric light bulb. (Courtesy of the University of Toledo's Ward Canaday Center.)

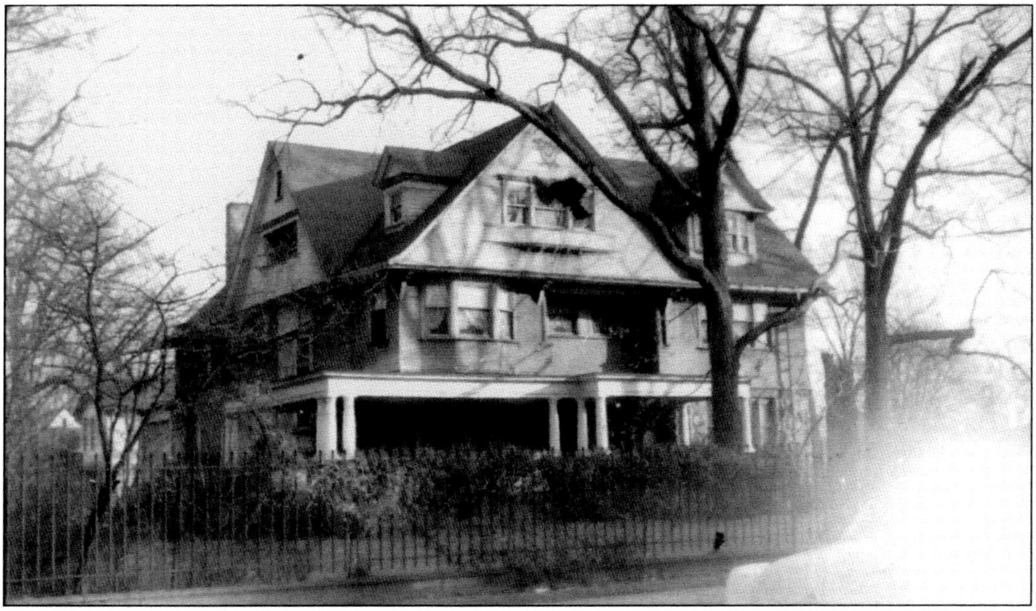

On his first trip to Toledo to discuss moving his glass company there, Edward Libbey met Florence Scott. On June 24, 1891, Libbey married Florence Scott of Toledo. As a single bachelor, Libbey lived at the Madison Hotel, from which he made his daily trip to the post office to pick up incoming receipts. The Scotts were a wealthy Lucas County pioneer family.

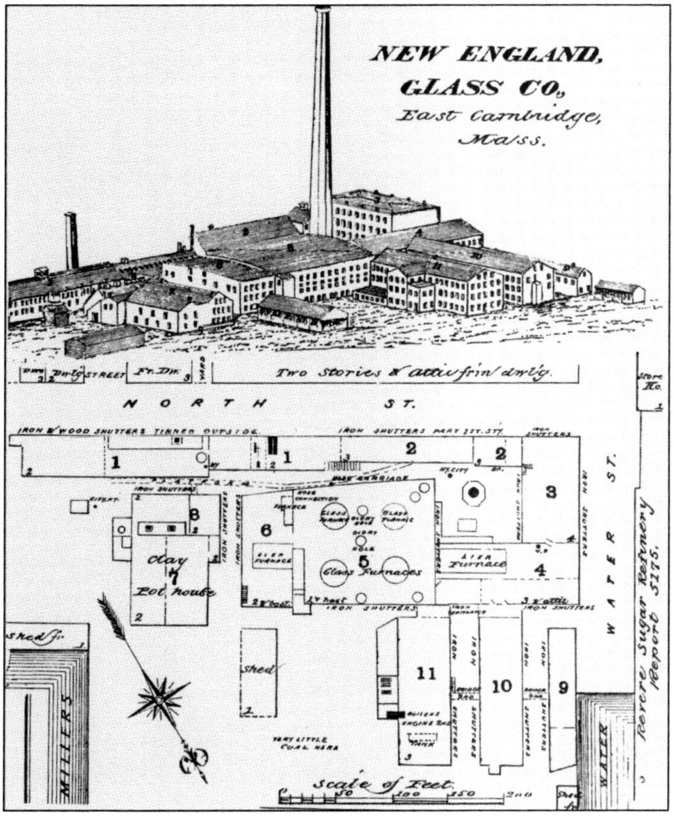

The New England Glass towering chimney was a Boston/Cambridge landmark in 1880. It had also built a reputation for cut crystal glass, taking many awards at the 1876 Philadelphia Centennial Exposition.

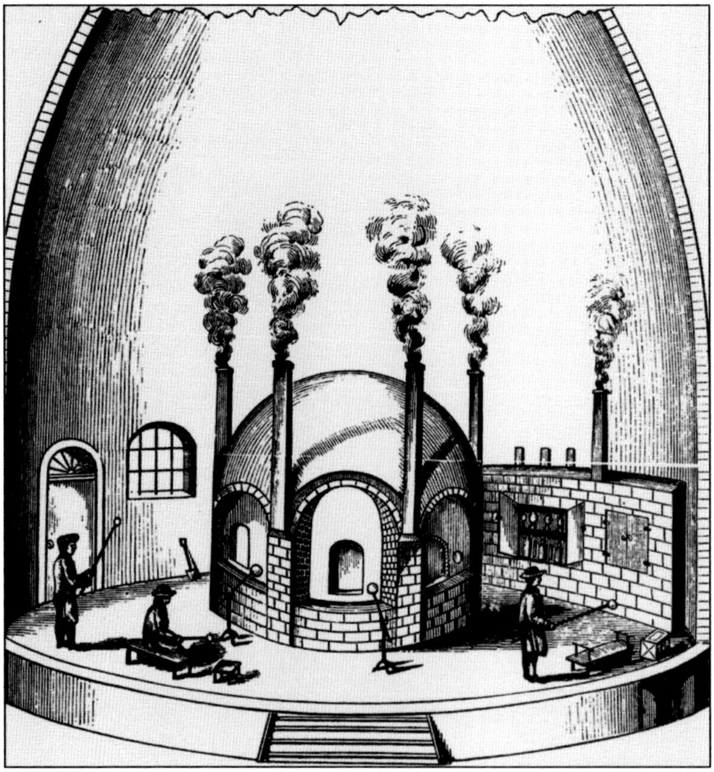

All the great cut glass makers of the world had adopted the Venetian glassmaking method. Note the benches or "chairs," which remained part of the glassmaker's setup into the 20th century. Also note the high chimneys, which were needed for wood furnaces to provide the necessary updraft.

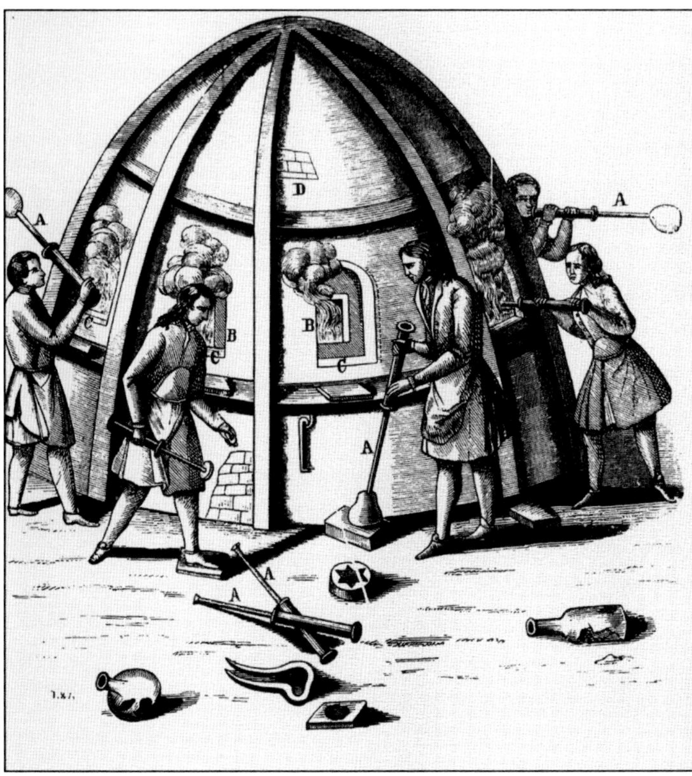

Note the use of wooden molds in bottle and container glassblowing.

Edward Libbey worked his way up through his father's company, but it was in sales that he excelled. His sales effort of Libbey's ruby red glass, known as Amberina, in 1883 turned the struggling company around. He also established a key customer in Tiffany's of New York for Amberina.

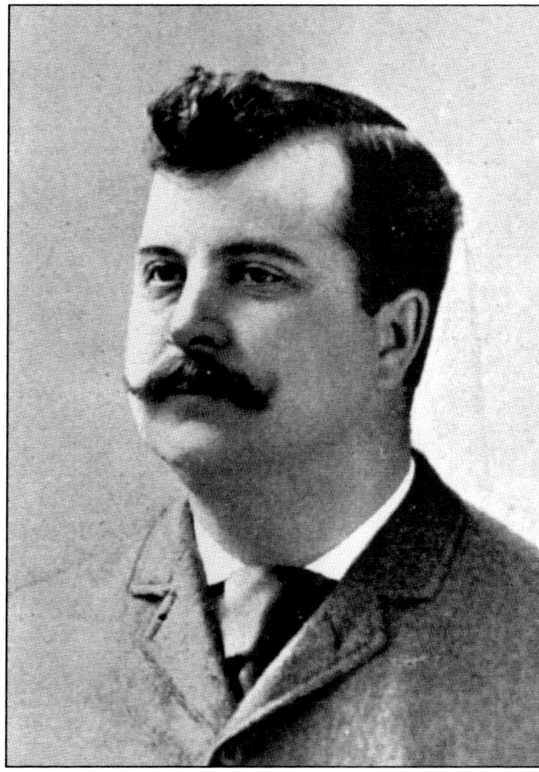

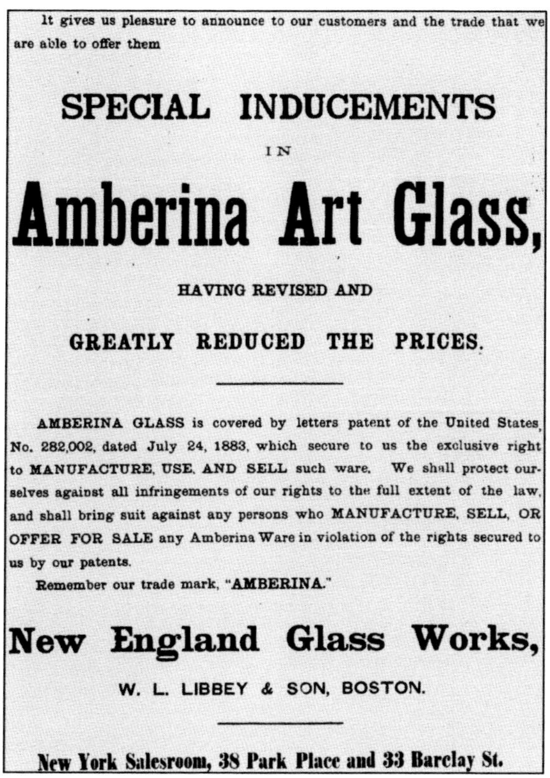

Libbey Company rediscovered the ancient technique of adding gold to glass to obtain rich blends of red. The process was patented in 1883 and fiercely protected by Libbey. Sales had peaked by 1888, and competitors had discovered the process.

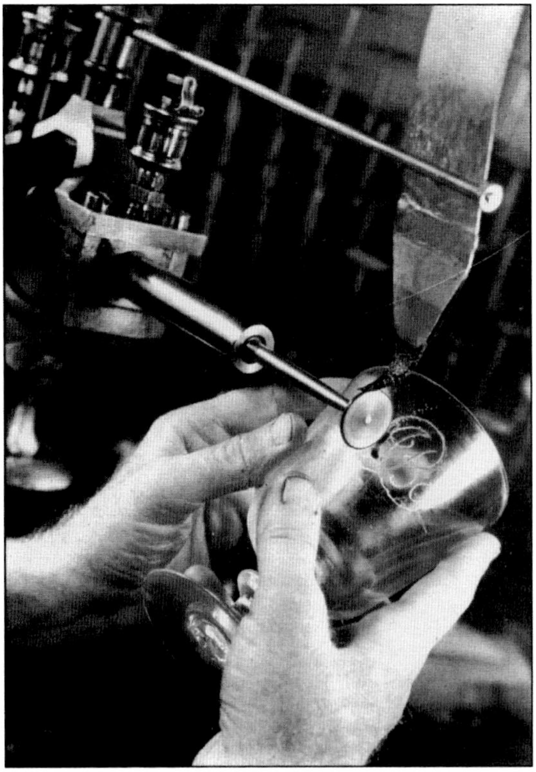

This was the most difficult of the glassmaker's handicrafts. A copper wheel was pressed against the glass piece as a leather tip fed a mix of linseed oil and emery powder. (Courtesy of the University of Toledo's Ward Canaday Center.)

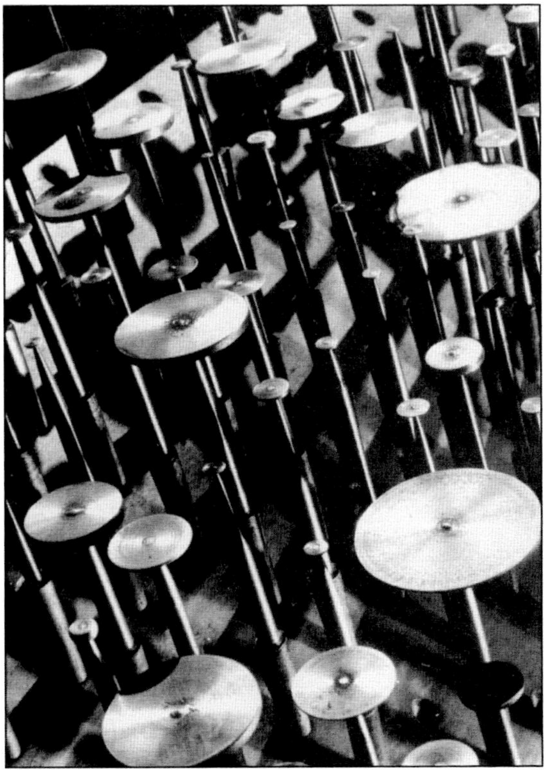

The glasscutter had a wide variety of sizes of copper wheels available, but great judgment was needed in their selection. One design might take as many as 50 different sizes. Libbey had the best copper wheel craftsmen in world.

Abrasive cutting used sandstone and in the 1900s, Carborundum, to make deep V cuts in a pattern.

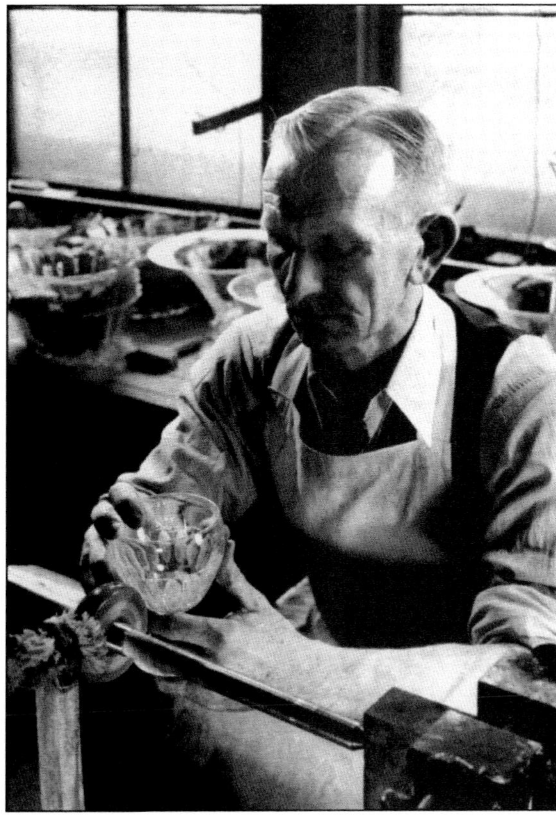

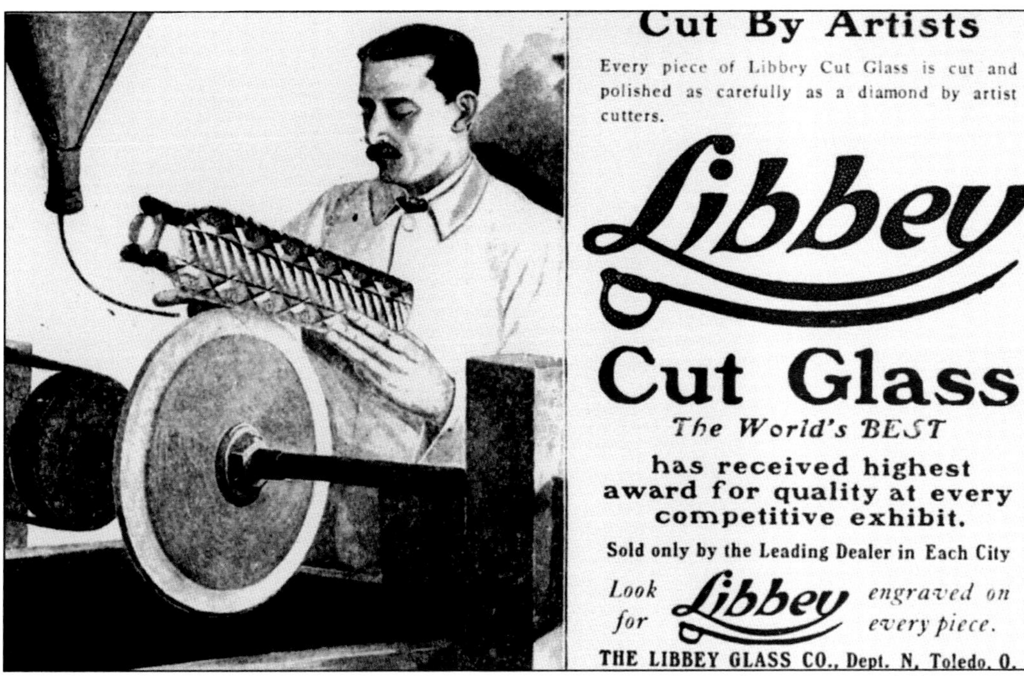

Libbey Glass would dominate the cut glass market for many years with its renowned glasscutters. Libbey Glass consistently took awards at world fairs.

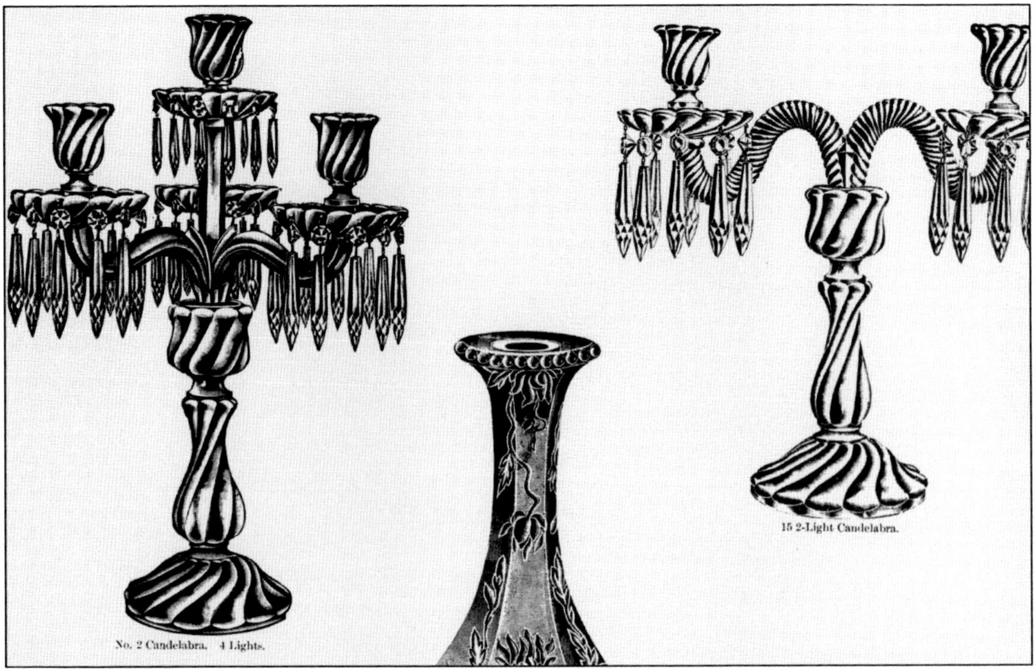

The cut glass market of lead crystal had an upscale market such as candelabras.

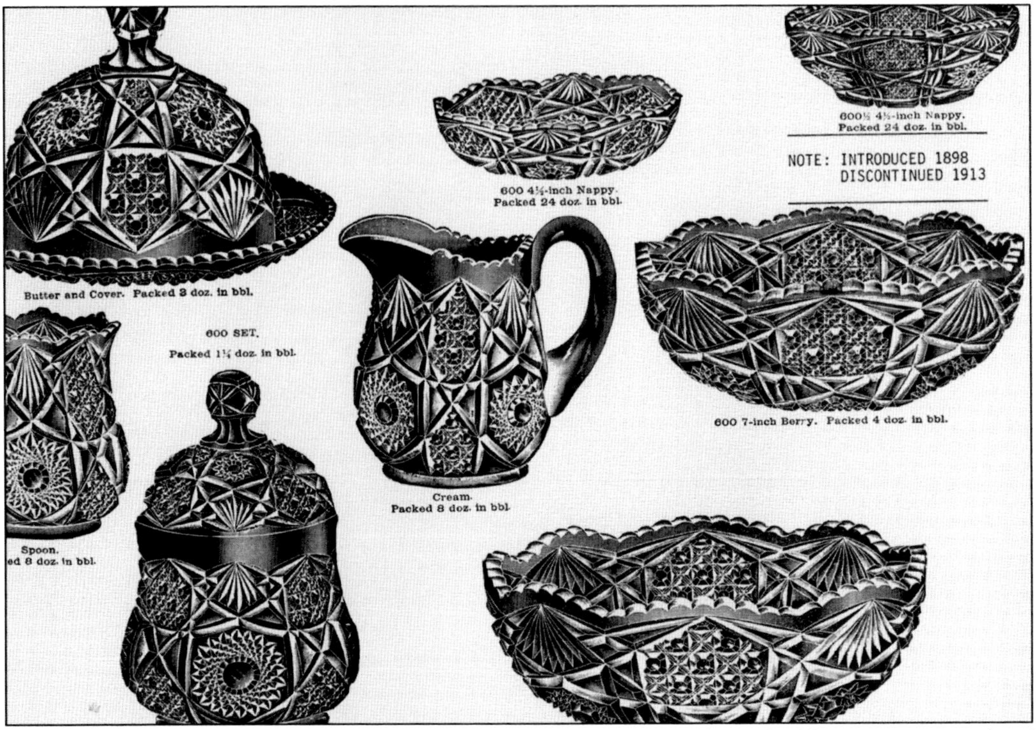

Pressed glass represented a much cheaper option for the average household. It was not made of heavy lead crystal but lime-soda glass. Ex-New England glassmaker William Leighton, who was brought to the West Virginia industry, developed pressed glass. The Brazilian pattern shown here was one of the most popular of 1899 by northwest Ohio's Fostoria Glass.

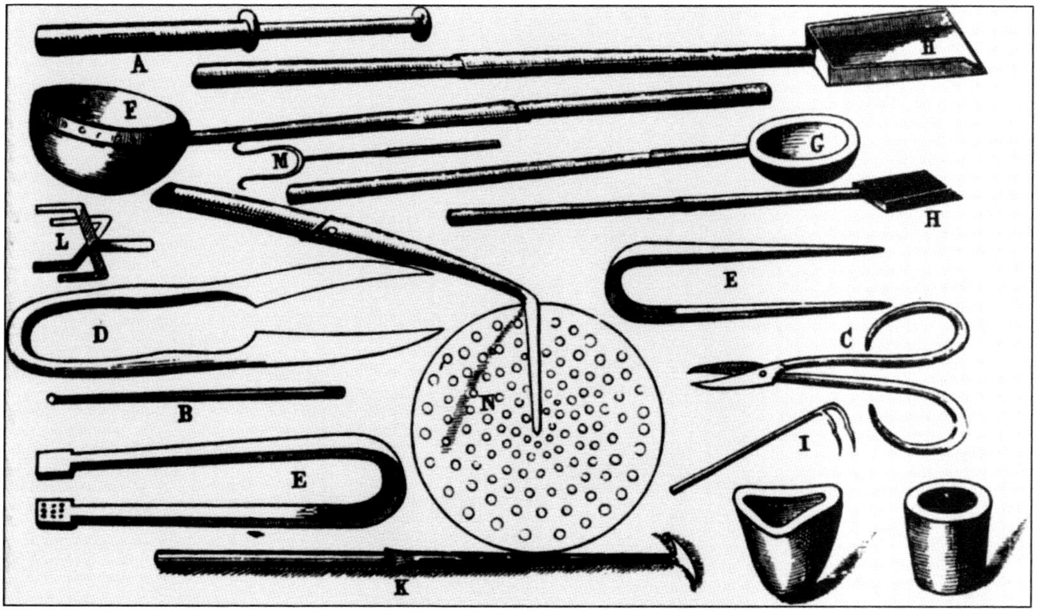

The Venetian glassmaking tools of the 12th century hardly changed in 800 years until the arrival of Mike Owens.

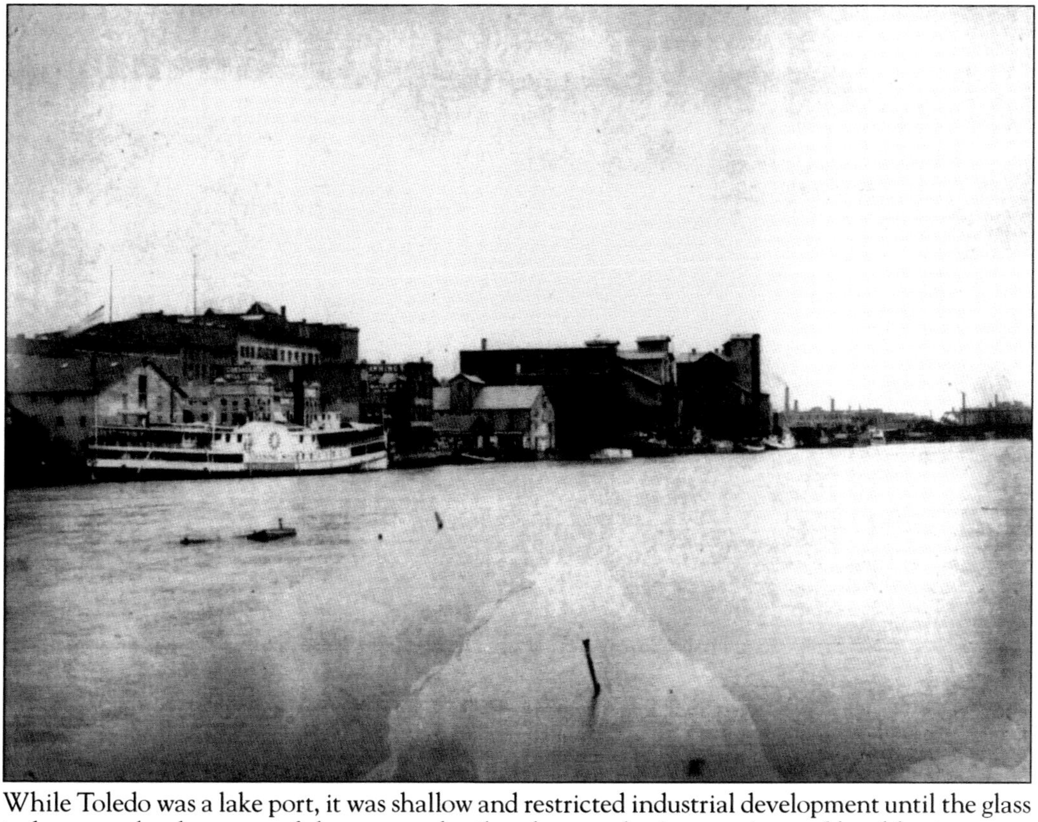

While Toledo was a lake port, it was shallow and restricted industrial development until the glass industry took advantage of the port and railroad network. A committee of local businessmen had been formed early in the 1880s to bring new industries to Toledo.

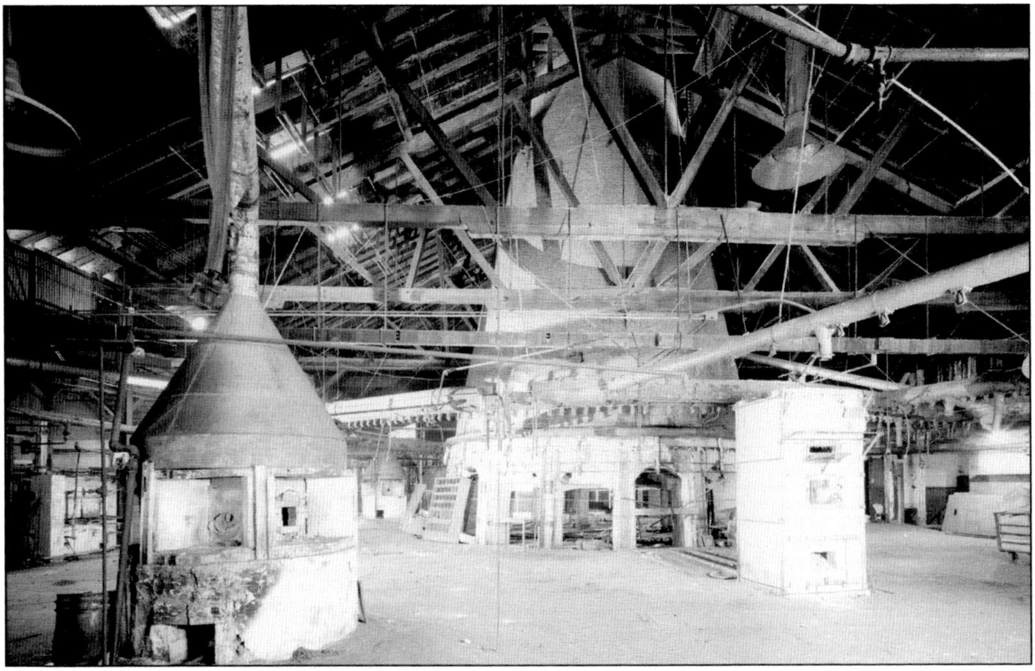

The layout of the typical 19th-century glassworks was not much different from those of Venice in the 16th century. One large wood- or coal-fired furnace heated 4 to 12 pots of glass. The small side furnace was an annealing furnace to remove work stress in the finished glass pieces.

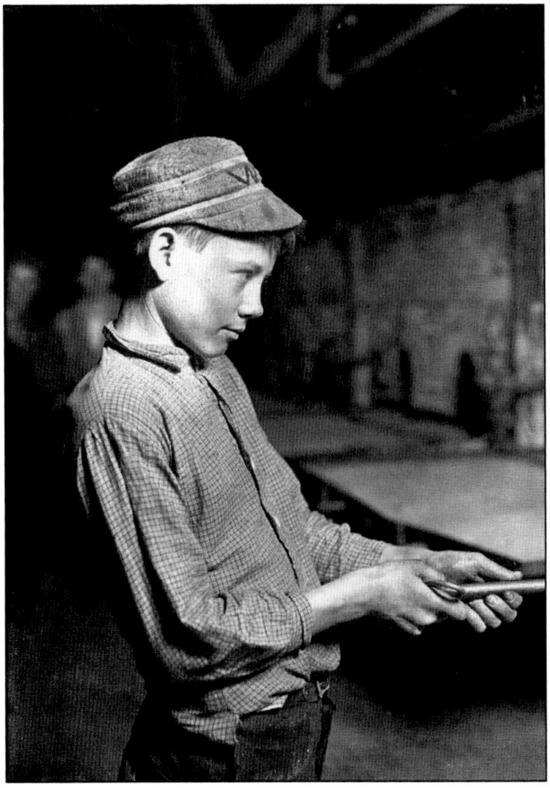

Approximately 25 to 40 percent of glasshouse labor was children. Boys were used to handle, move, clean, and anneal glass pieces. They were also employed to maintain furnaces and package final product. Boys from 8 to 12 years of age were commonly employed regardless of state laws.

Two

Gas Lights the Way to Northwest Ohio

On January 20, 1886, at present-day Liberty Street (in downtown Findlay), a major gas strike was made. The find resulted in a gas boom and industrial boom for northwest Ohio, which had no glasshouses in 1886. By the end of 1887, there were 10 glass factories in Findlay. At the same time, Bowling Green attracted seven large glassmakers and was dubbed the "Crystal City." Unfortunately for Toledo, the Trenton deposit did not reach to the city, but the Toledo Businessmen's Association was an innovative brunch. A pipeline was built in 1887 to Toledo.

The first glass company to come to northwest Ohio was Findlay Window Glass. Within a year, it was followed by four other window glass companies—Wilcox Glass, Buckeye Window Glass, Ohio Window Glass, and Hirsch-Ely Window Glass. By late 1887, the more established art, houseware, and bottle glass companies were moving to northwest Ohio from Pittsburgh and West Virginia. By 1889, Findlay was the second-largest glassmaking center, second only to Pittsburgh. The number of glasshouses in a few years went from 0 to 25 in Findlay, 11 in Bowling Green, 15 in Fostoria, 5 in North Baltimore, 4 in Maumee, 8 in Tiffin, 8 in Toledo, and another 15 or so scattered in smaller communities. Within a few years, northwest Ohio had many more glasshouses than Pittsburgh, New England, and West Virginia.

The exponential growth of glasshouses caused a critical shortage of labor, in particular, child labor, which was essential to glasshouses, making up most of the employees. Glasshouses could not function without mold boys, carry-in boys, carryout boys, and general laborers. Orphanages all over the United States became sources of supply. There were always rumors of "orphan trains" arriving throughout the night in northwest Ohio train stations. From all over the country, boys who had minor problems in probate court were shipped to the glasshouses. At many glasshouses, such as Nickel Plate Glass in Fostoria, these kids put in their 10 hours a day in two shifts, one shift from 7:00 a.m. to noon and another from 5:00 p.m. to 10:00 p.m.

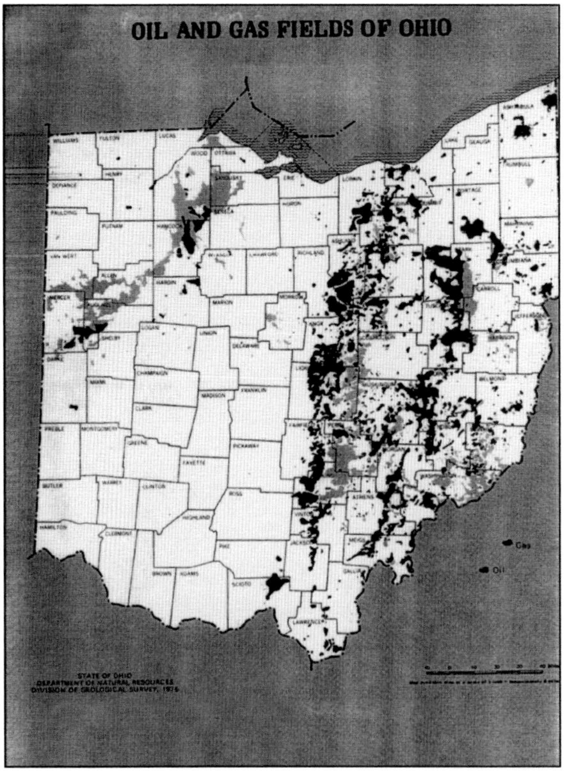

Early settlers of northwest Ohio had reported the illumination of swamp gas in the early 1800s. An oil strike in Lima in 1885 heralded the discovery of the Trenton limestone strata, rich in oil and gas. The gas deposits were centered in Hancock County. Notice that part of the Trenton deposit drifted into Indiana and was a basis for Indiana's glass industry. (Courtesy of the Bowling Green State University Center for Archival Collections.)

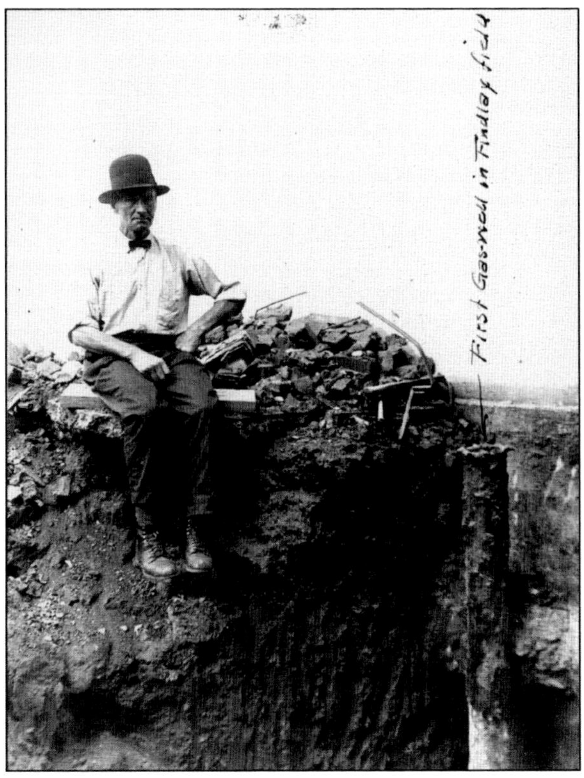

Dr. Charles Osterlen drilled several farm wells in the Findlay area in the 1880s. One Findlay farmer had been using natural gas as early as 1836. Dr. Osterlen used his knowledge of geology to find highly productive wells. (Courtesy of the Hancock Historical Society.)

On January 20, 1886, the Karg well struck gas in the middle of the village of Findlay. The location was on the south bank of the Blanchard River on Liberty Street. The light of the well could be seen as far as Toledo. The noise was deafening as 12 million cubic feet fed the flame every day. The well struck high-pressure gas at 1,000 feet in what was known as the Trenton limestone strata. Amateur geologist Charles Osterlen had found gas on his farm in 1884 prior to tapping the great Karg strike. (Courtesy of the Hancock Historical Society.)

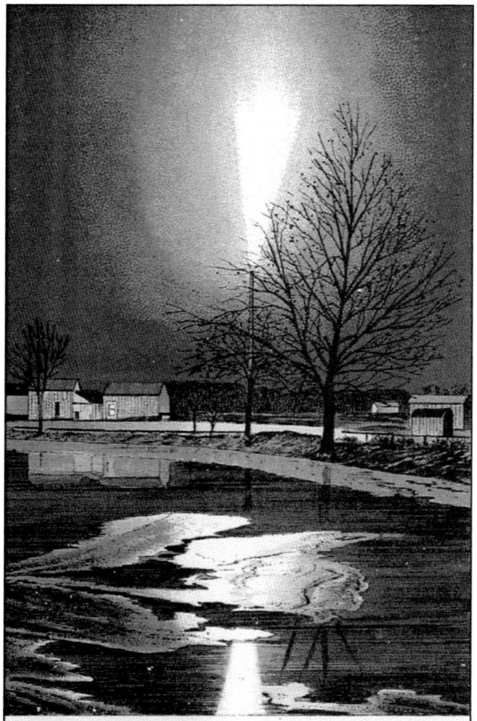

THE GREAT KARG GAS WELL,
SEEN AT NIGHT FROM MAIN ST. BRIDGE PRESSURE 800 L B S.

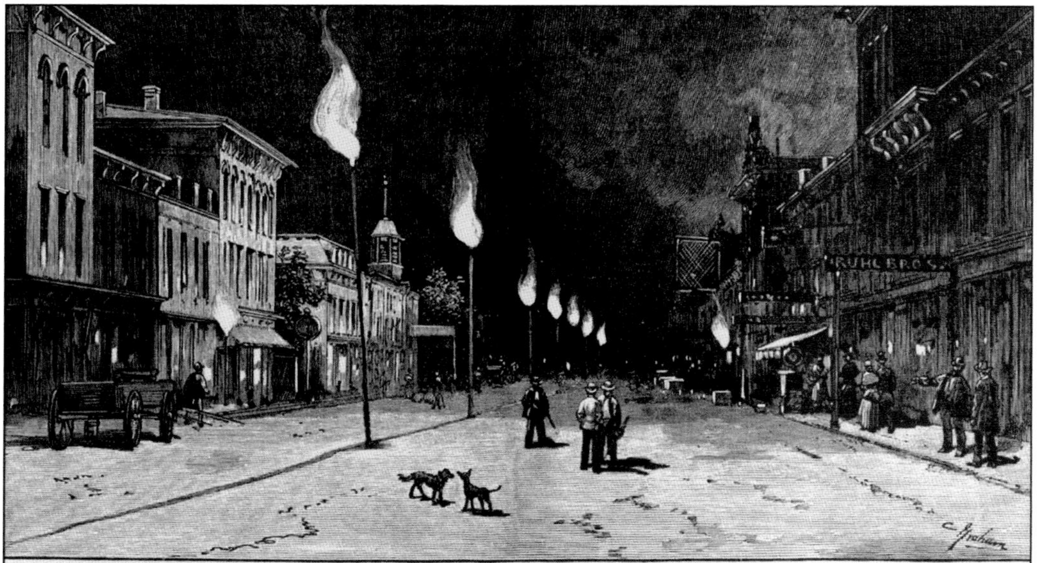

"EET LIGHTED BY NATURAL GAS AT FINDLAY, OHIO.—DRAWN BY CHARLES GRAHAM FROM PHOTOGRAPHS BY ZAY, AND CROZIER & LINAWEAVER, FINDLAY, OHIO.

Within months of the Karg gas strike, gas was being supplied to retailers at a cost of $3 a month. Gas lights for street, store, and homes burnt 24 hours a day (there were no pilot lights, valves, or meters). The chamber of commerce ran national advertisements offering free natural gas and land for factories to move to Findlay. By 1888, Findlay had attracted 72 factories, including 11 glasshouses. By 1891, demand outstripped supply, and by 1895, the low gas pressure and dry wells ended the gas boom in northwest Ohio. (Courtesy of the Hancock Historical Society.)

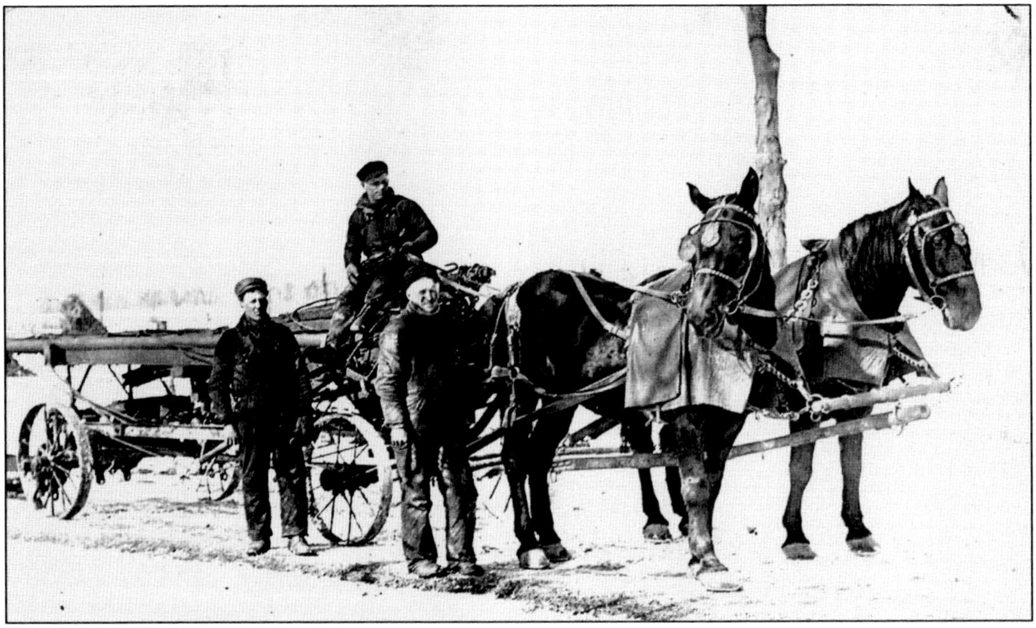

In 1887, two rival Toledo companies, Toledo Natural Gas and Northwestern Gas brought up Findlay wells, proposing a pipeline to Toledo. Construction of the lines started in the winter of 1887. On July 27, 1887, former president Rutherford B. Hayes opened the lines at Toledo's Grand Gas Celebration. The gas lines followed closely present-day I-75. The pressure in the pipelines dropped so low in the cold winter of 1888 that Libbey Glass's pots had to be warmed by cutting wood. In December 1888, gas shortages caused hundreds of homes to go cold for days. (Courtesy of the Bowling Green State University Center for Archival Collections.)

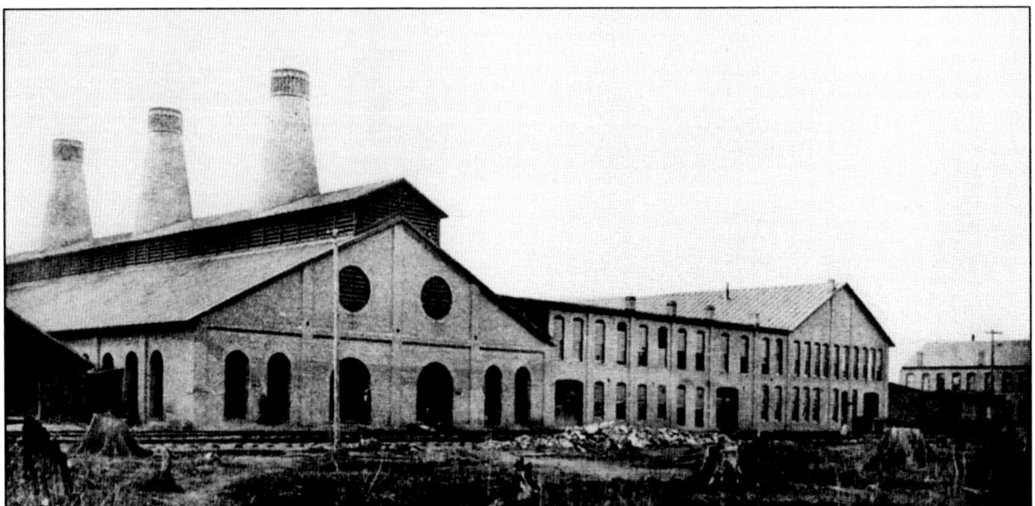

The Dalzell Glassworks was one of the first glass companies and Findlay's longest-lived glass company, closing in 1901. The Dalzell brothers were glass men from Wellsburg. Edward Gilmore was a Pittsburgh banker, and William Leighton Jr. was the son of the famous New England Glass Company chemist William Leighton, who invented pressed soda glass. The company produced crystal flint glass for lamps and fine tableware. Like his father, William Jr. helped Dalzell Glass become a leader in colored glass.

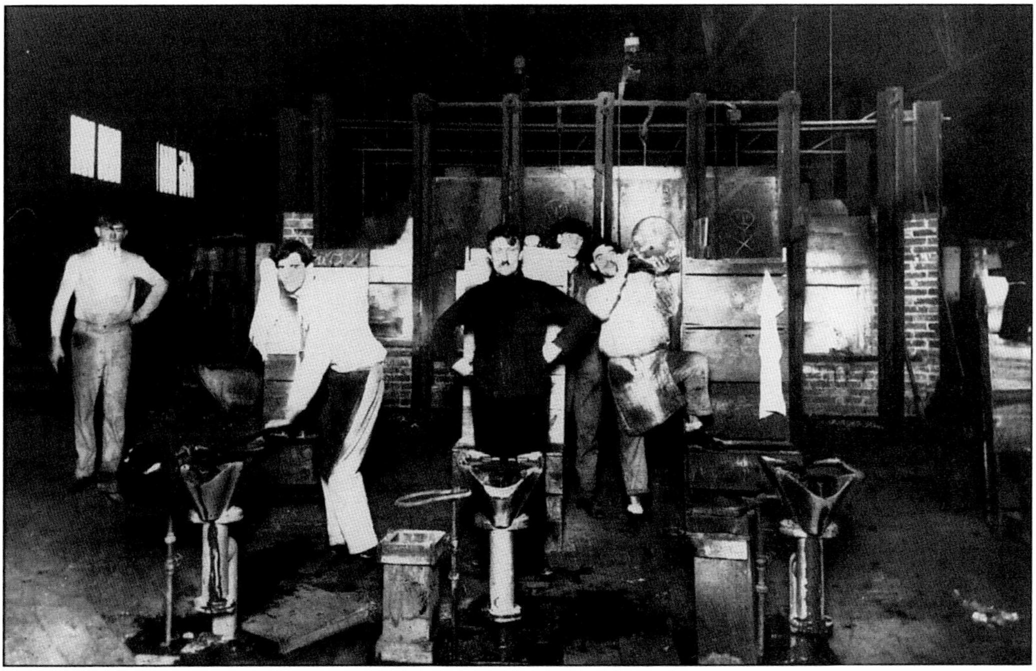

The strange saddlelike molds were used for pressed glass manufacture. The great glass chemist William Leighton came to run the operation from Hobbs-Brockunier of West Virginia (his father had been an expert in colored glass at New England Glass). The fine colored tableware of this operation was world famous. (Courtesy of the Hancock Historical Society.)

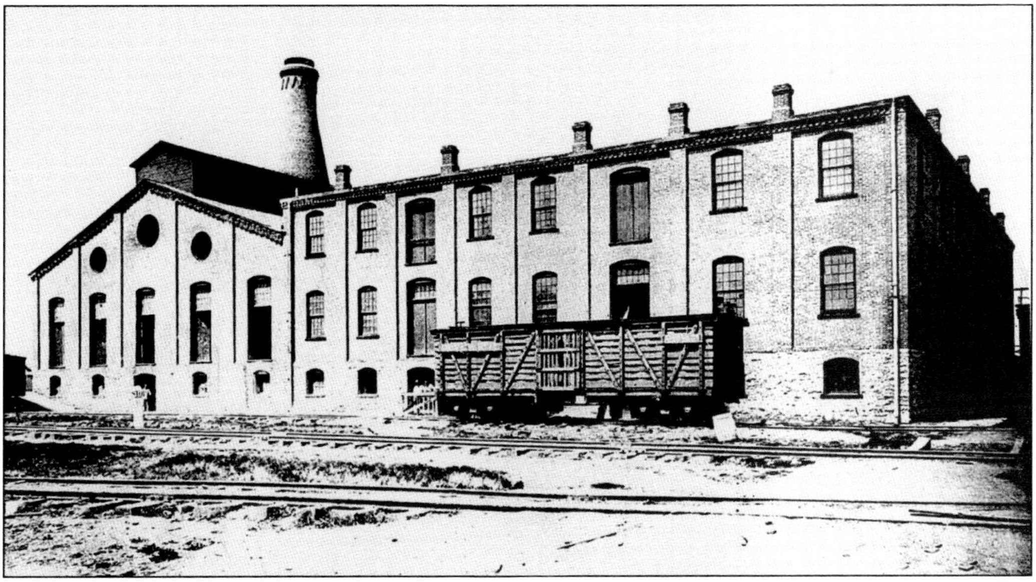

Columbia Glass was the first tableware manufacturer in northwest Ohio, opening in Findlay in 1886. Formed by Pittsburgh glassmakers, Columbia was a major manufacturer of blown and pressed glass. David Jenkins, the president of Columbia, would form America's first "glass trust," consisting of 17 companies throughout the United States. In 1890, the Findlay plant became known as United States Glass Company, Factory J. Due to the gas shortage, the company tried to switch to oil but ultimately failed in 1893.

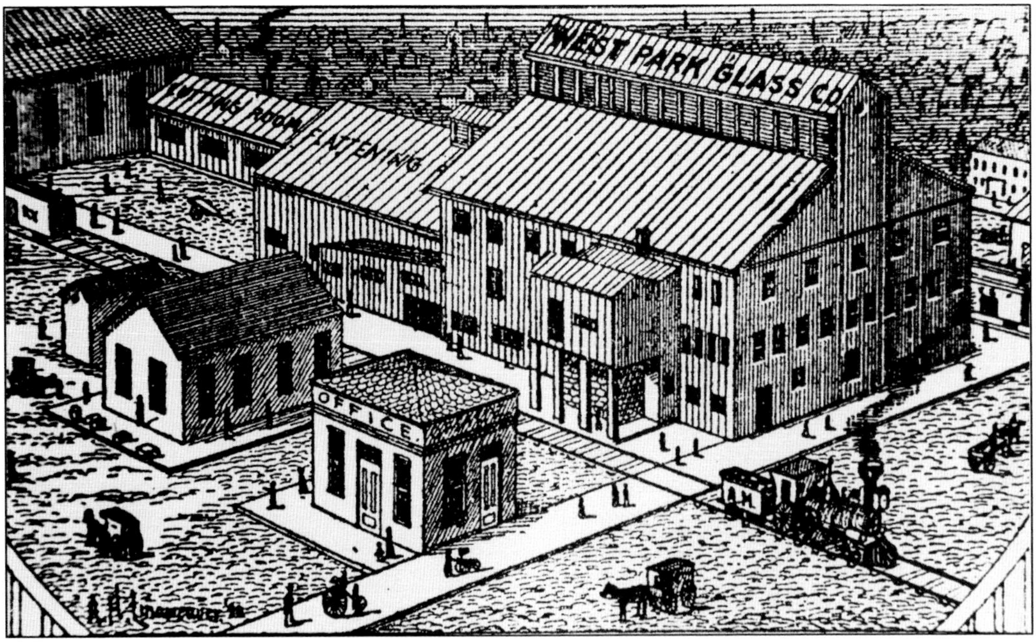

This was one of six window glass companies that came to Findlay for free natural gas. It was built at a railroad intersection near Race Street and Morrical Avenue to facilitate product shipping. One of the many "mobile wooden structures" in 1896 when the gas failed, it packed up 25 railcars and moved to Indiana. (Courtesy of the Kaubisch Memorial Public Library.)

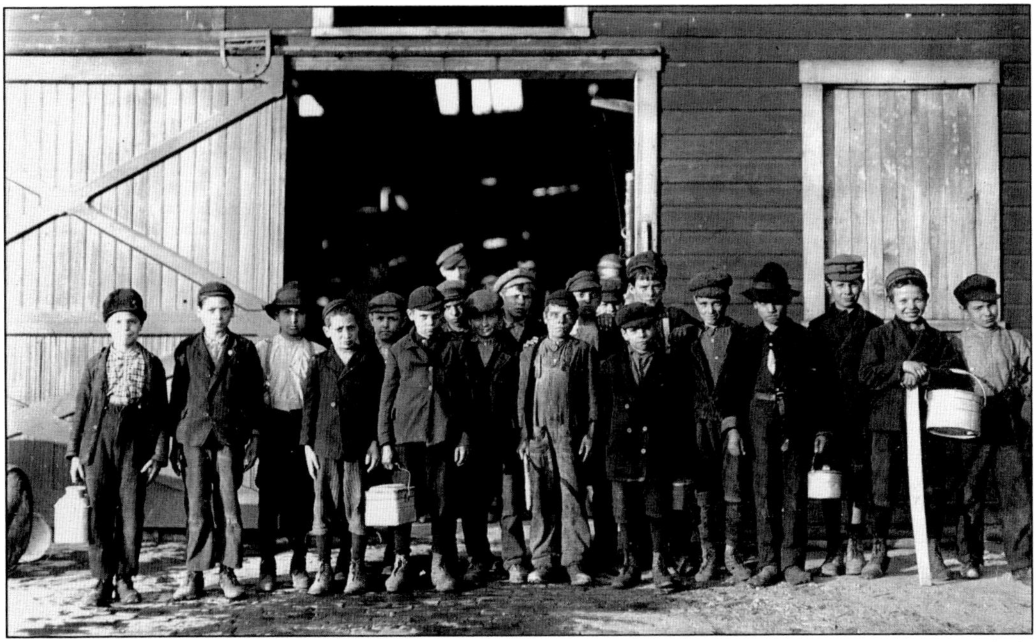

The glass boom of the 1880s created even more abuses in the employment of children. While Ohio was one of the few states that had child labor laws, these laws were typically ignored by the glass industry. Much of the justification was based on the use of orphans, thus providing them a career. Boys leaving the orphanage were given "two suits, four pairs of socks, a good pair of shoes and a hat." Estimates suggest boys made up about 30 to 40 percent of the workforce.

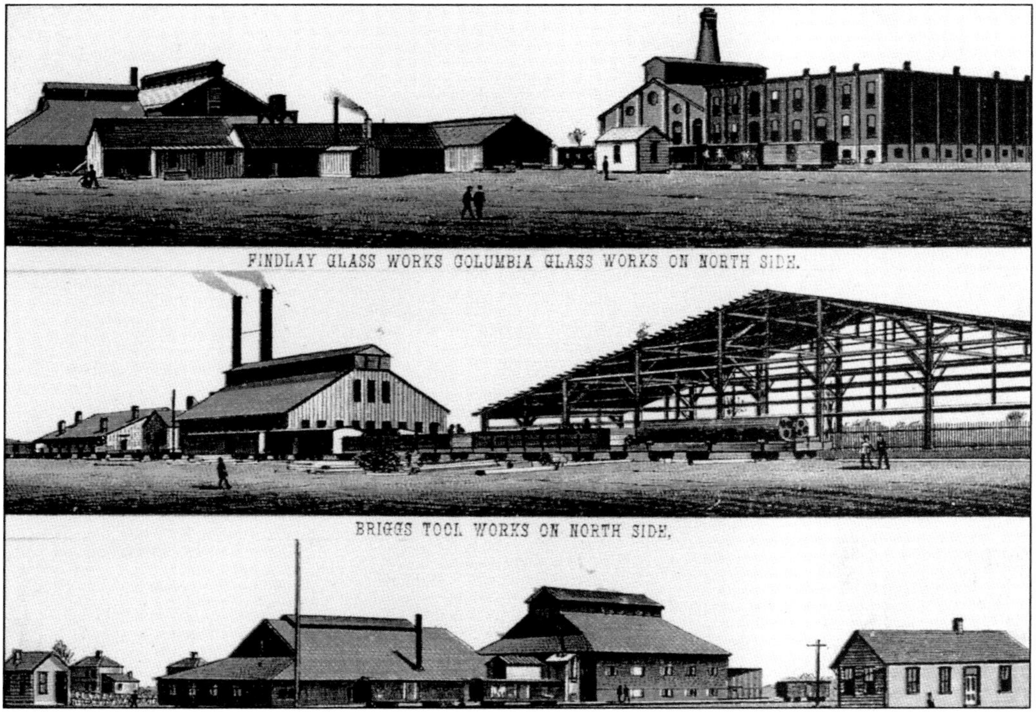

This was the first window company to build in Findlay in 1887. It was a wooden structure built quickly, and its eight-pot furnace operated less than a year.

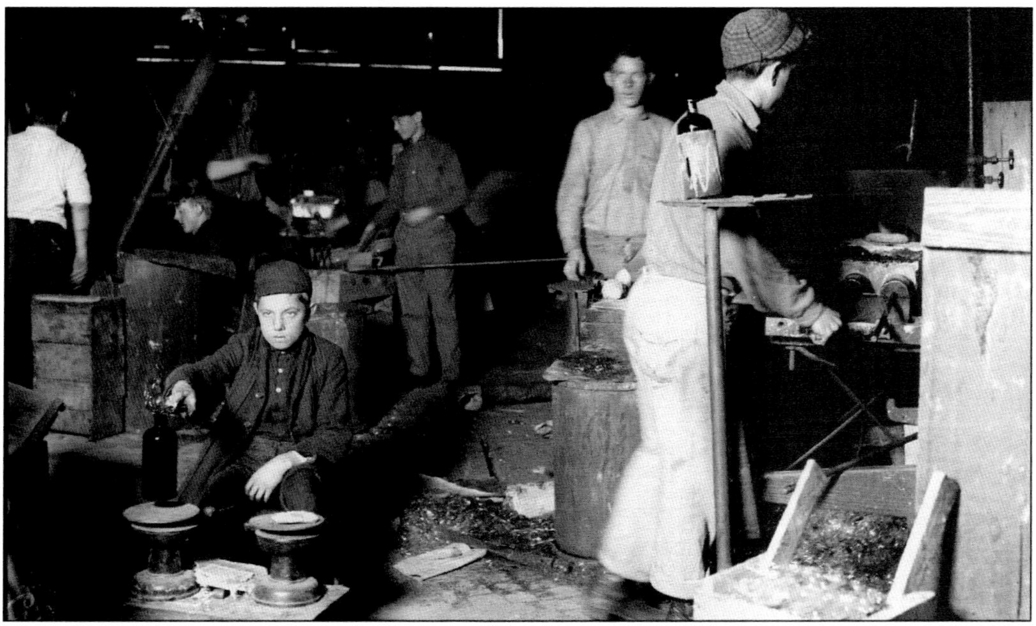

North Baltimore, Ohio, was a railroad junction of the Baltimore and Ohio with the Toledo, Columbus, and Southern railroad. Located north of Findlay in the Trenton gas field, it drew some of northwest Ohio's first glass works. North Baltimore ran until 1900, when the low gas pressure forced it to pack up and move to Terre Haute, Indiana. (Courtesy of the Bowling Green State University Center for Archival Collections.)

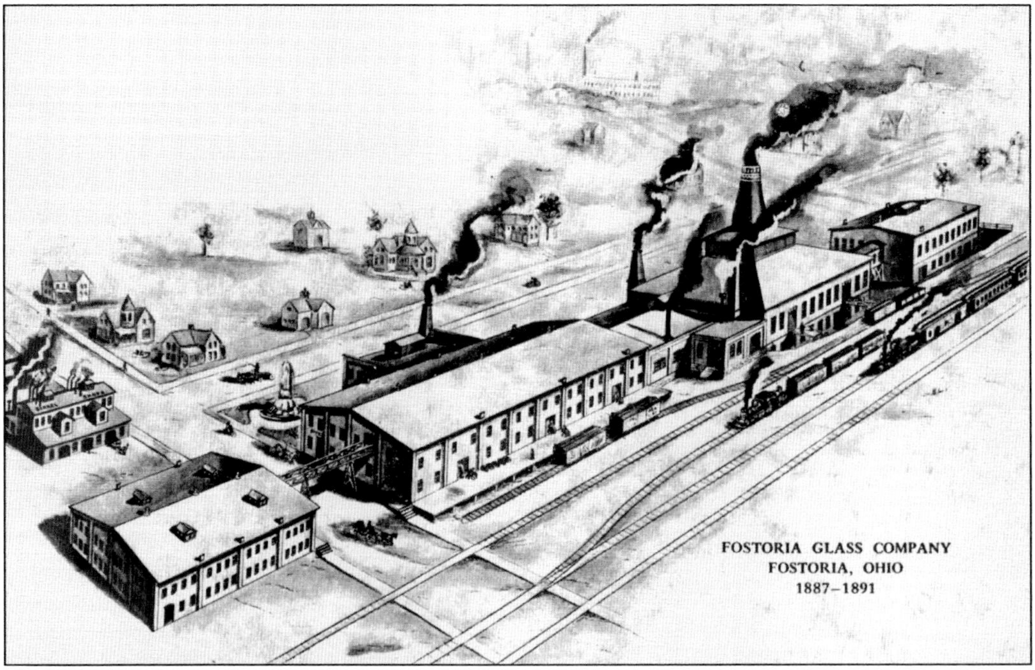

Fostoria Glass was one of the earliest companies to come to northwest Ohio in 1887. The 12-pot furnace was located on South Vine Street in Fostoria. The city gave gas to the plant for $200 a year and offered a $10,000 signing bonus. The company left in 1891 when the gas ran out.

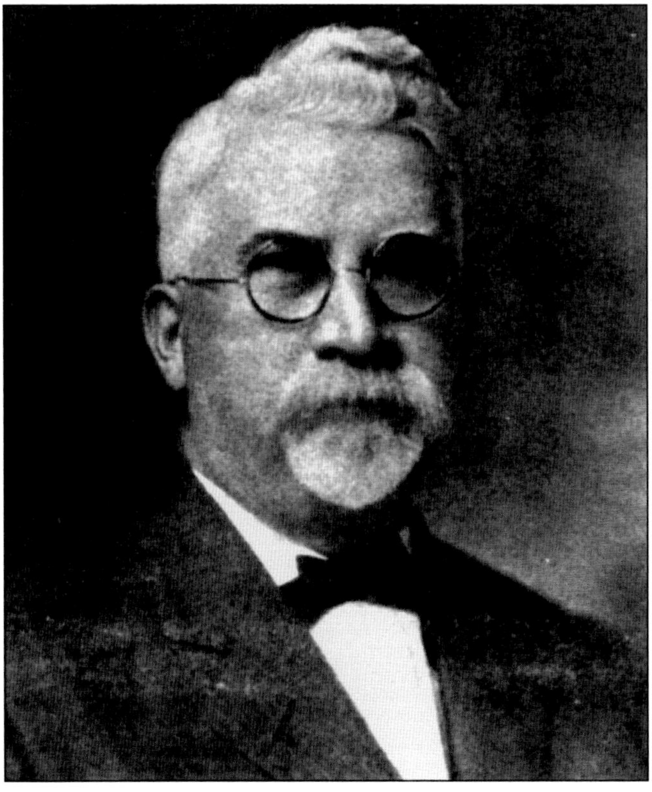

William Dalzell brought glassworks to Findlay and Fostoria in 1887. Dalzell was an early pioneer of automated glassmaking even before Owens and Libbey. At a Findlay hotel in 1901, he formed National Glass, a major glass trust of companies. The trust failed, and Fostoria Glass moved to Moundsville, West Virginia, in 1891.

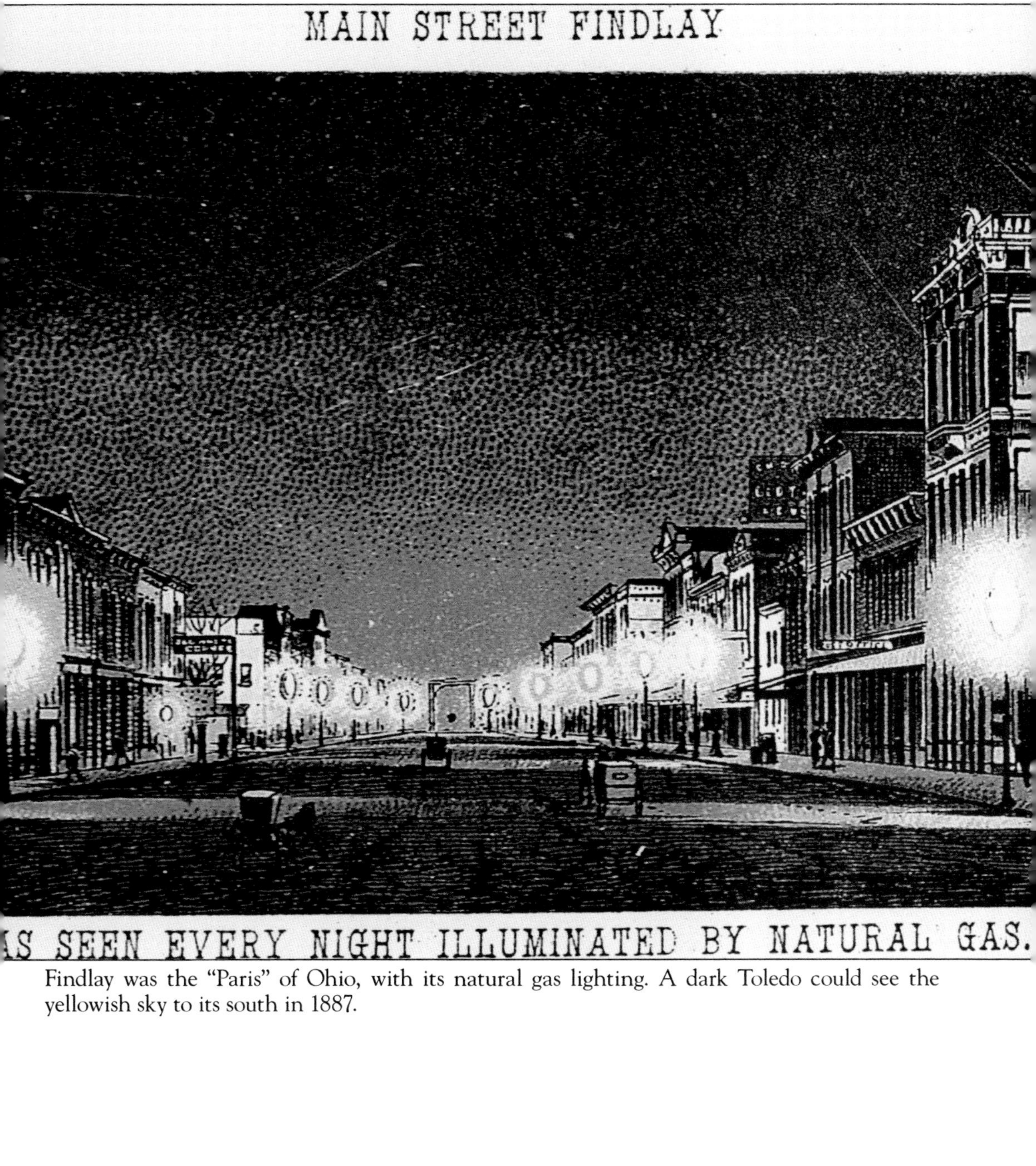

Findlay was the "Paris" of Ohio, with its natural gas lighting. A dark Toledo could see the yellowish sky to its south in 1887.

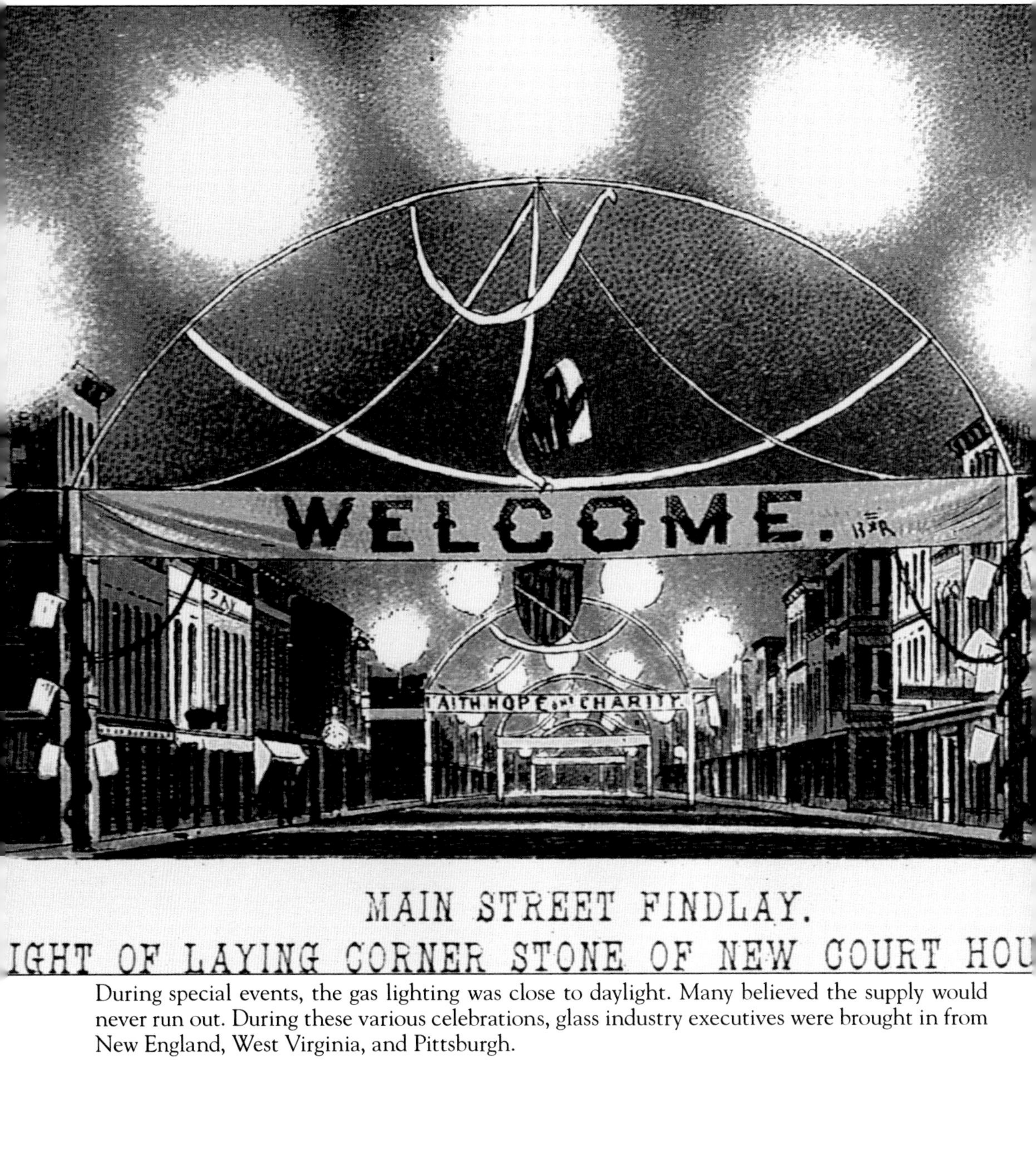

During special events, the gas lighting was close to daylight. Many believed the supply would never run out. During these various celebrations, glass industry executives were brought in from New England, West Virginia, and Pittsburgh.

Three

LIBBEY SELECTS TOLEDO

The New England Glass Company was suffering from what would be the first wave of globalization in the glass industry. The new owner, Edward Libbey, following William's death, had no choice but to look west for cheaper resources. The gas boom had brought Libbey to northwest Ohio, but he proved his business genius in the site selection. Libbey passed over offers for free land and gas from Lima, Findlay, and Tiffin. Libbey selected Toledo because it offered more than natural gas. Toledo assets included high-purity sand, rail connections, an Irish population for his workers' assimilation into the community, and a baseball team. Libbey knew the pain of abundance to shortage, and he planned in the flexibility of his new plant to use multiple fuels. And probably just as important was the extensive railroad network to the major western markets for his glass markets. Toledo offered coal at the price of that in Pennsylvania because of its railroad connections. Oil was also available from Cleveland pipelines.

In 1888, Toledo looked more like Venice than present-day Toledo. The great Ohio canal system of the 1850s connected Toledo to New Orleans via the Cincinnati-Toledo canal. The marshlike environment favored frogs—thus the early name of Frog Town.

A final stand by the New England union on wages left Libbey no choice but to leave his beloved Boston. Thanks to the persistence of the Toledo Businessmen's Association and Libbey's fascination with a young girl, Florence Scott, who would be the future Mrs. Libbey, Libbey choose Toledo.

Libbey chose to bring most of his Boston workforce with him. On August 18, 1888, a train loaded with Irish workers and 10 carloads of equipment arrived at Toledo's Union Depot. Toledo prepared with a large parade, bands, and picnic food. Ultimately it took a total of 50 railroad cars to transport New England Glass to Toledo. Within a few months, many homesick workers returned to Boston. Libbey himself suffered from homesickness as his first year resulted in heavy losses and an inability to get the plant running smoothly.

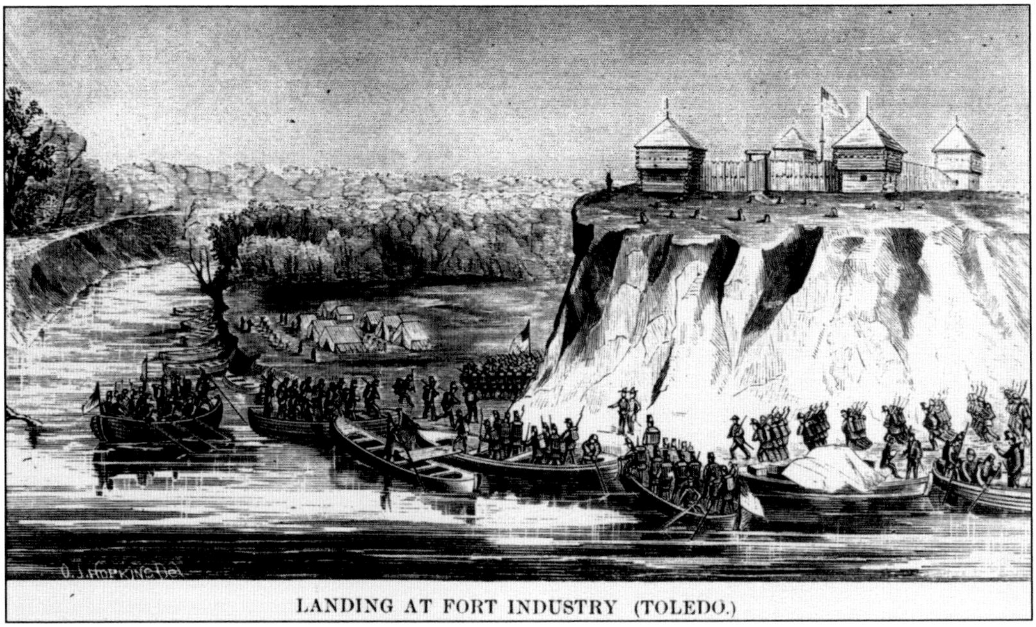

Fort Industry at the Maumee River was the first settlement of the Toledo area. Fort Industry offered high ground from the miles of swamp.

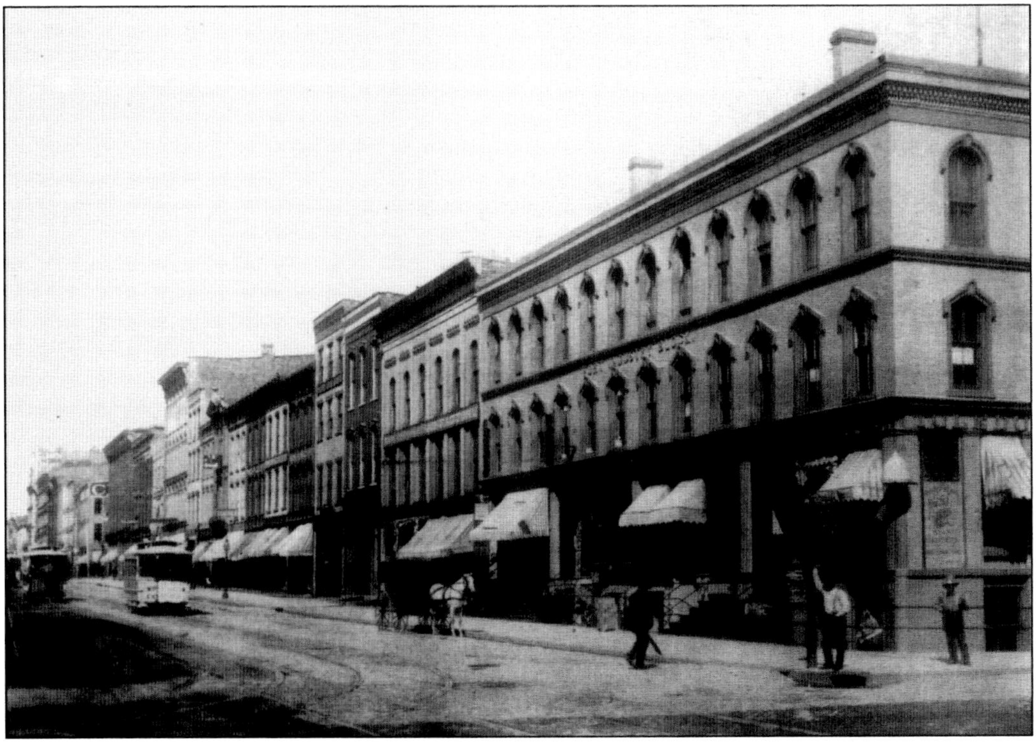

Seen here is Toledo in 1888, the year of Edward Libbey's arrival. Note the early streetcar system, which was extensive for a northwest Ohio city.

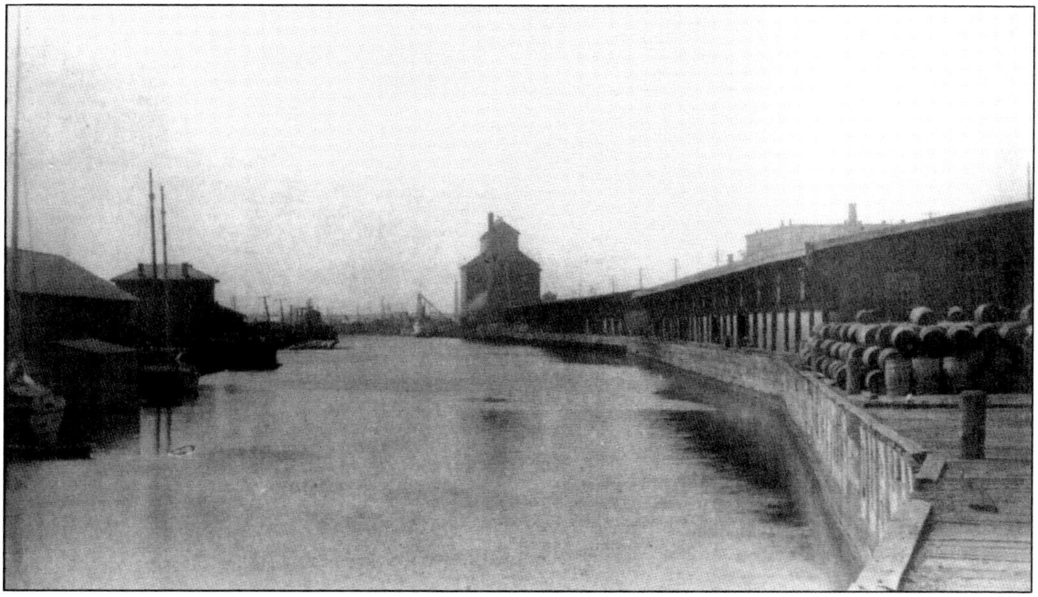

The city of Toledo was a network of wooden docks with wooden planks used as sidewalks in 1888. These paved avenues supported an electric trolley system, which even extended into the smaller burgs of northwest Ohio.

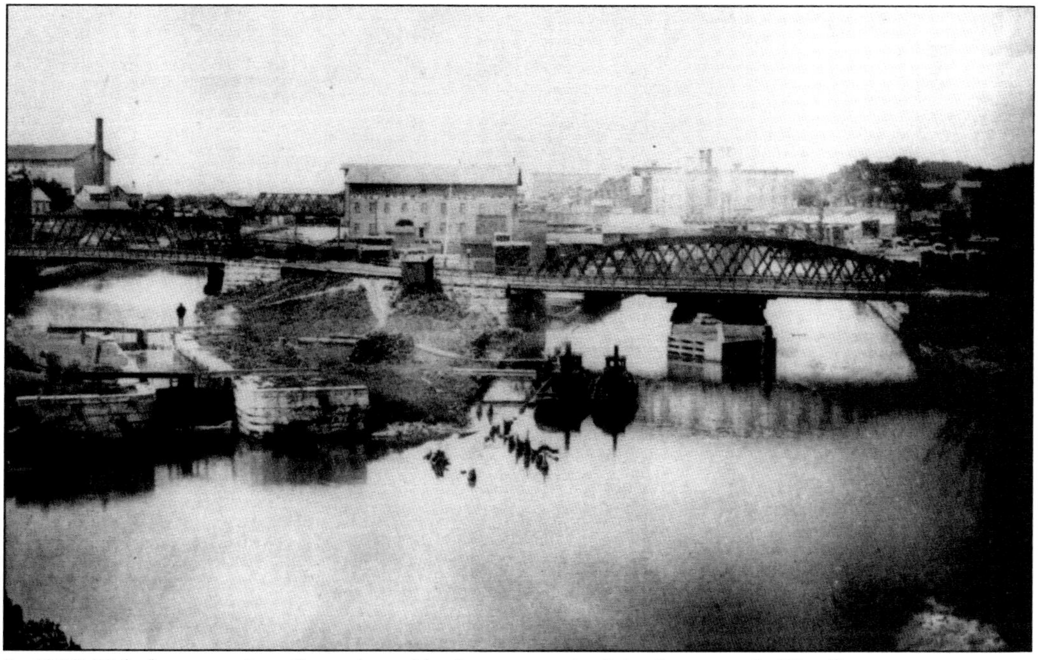

In 1888, Toledo was a city of canals and locks, as seen in this photograph. The Boston immigrants complained of the mosquitoes, humidity, and mud.

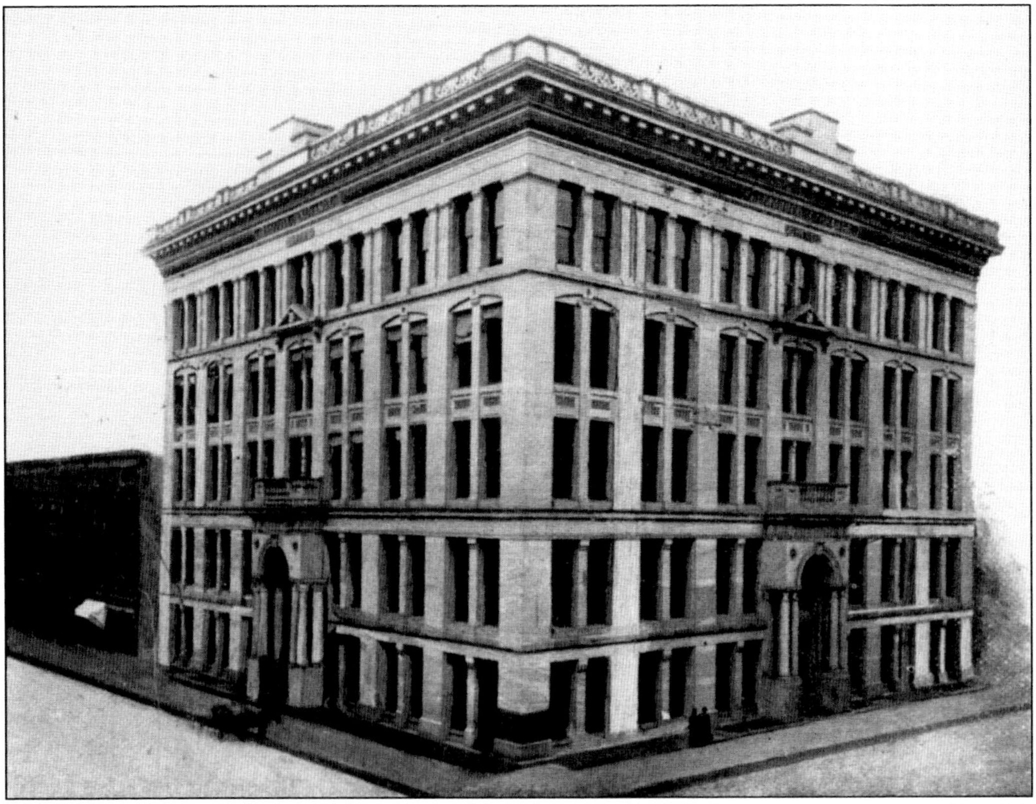

The Produce Exchange was the dominant building in 1888 Toledo. It was a tribute to Toledo's title of the corn capital of the world and clover capital of the world. In the 1870s, Toledo was even challenging Chicago as the future grain center.

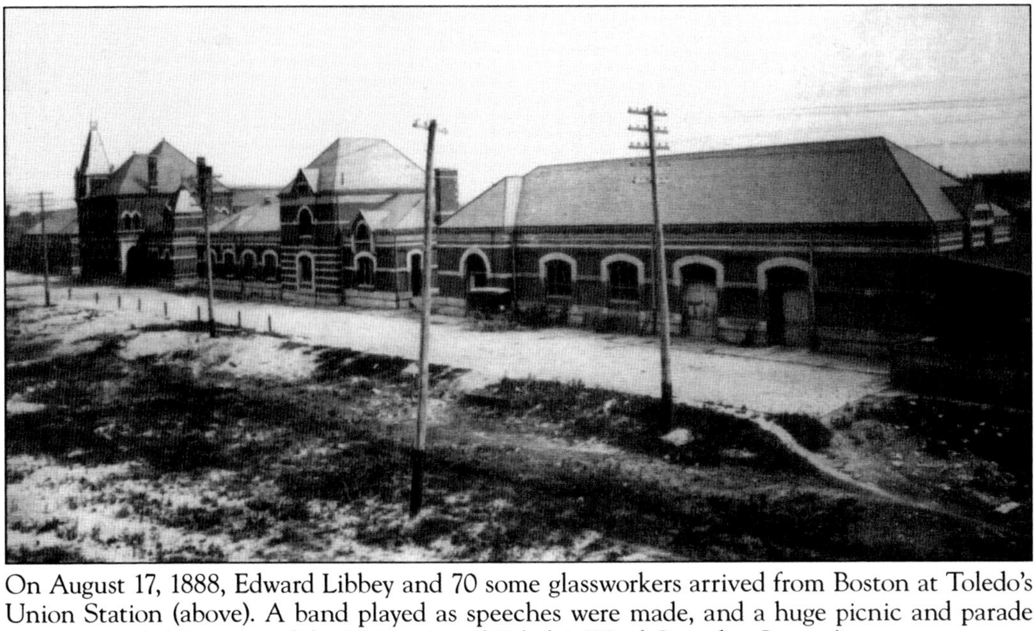

On August 17, 1888, Edward Libbey and 70 some glassworkers arrived from Boston at Toledo's Union Station (above). A band played as speeches were made, and a huge picnic and parade were planned. (Courtesy of the University of Toledo's Ward Canaday Center.)

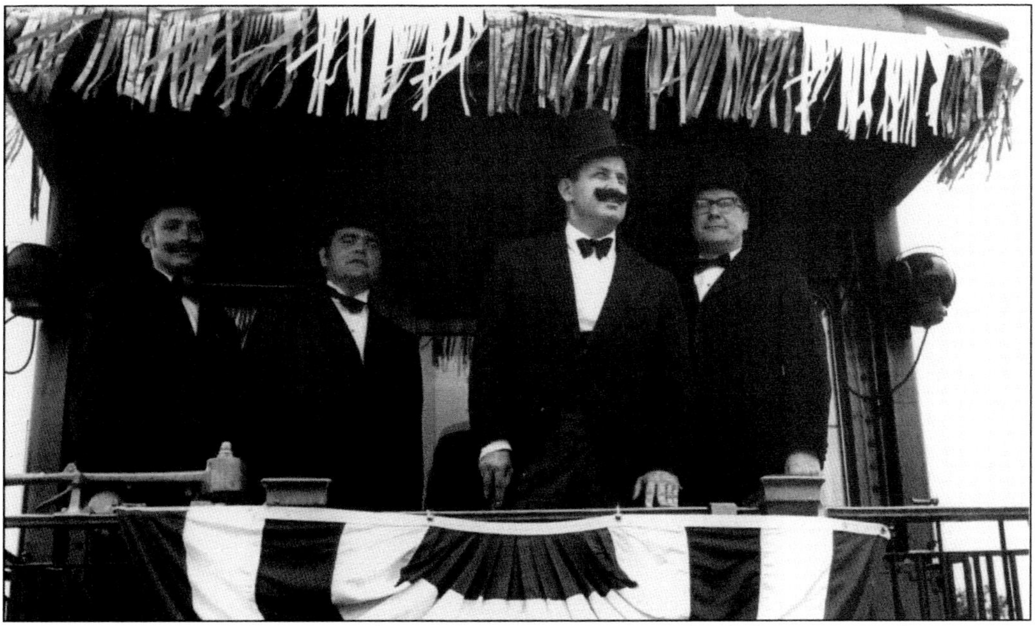

From left to right are M. Vogan playing Gen. James Kounts of Toledo, Mayor William Ensign playing Maj. Kent Hamilton, Ben Sanders playing Edward Libbey, and James Burg playing Sol Richardson at the Union Station.

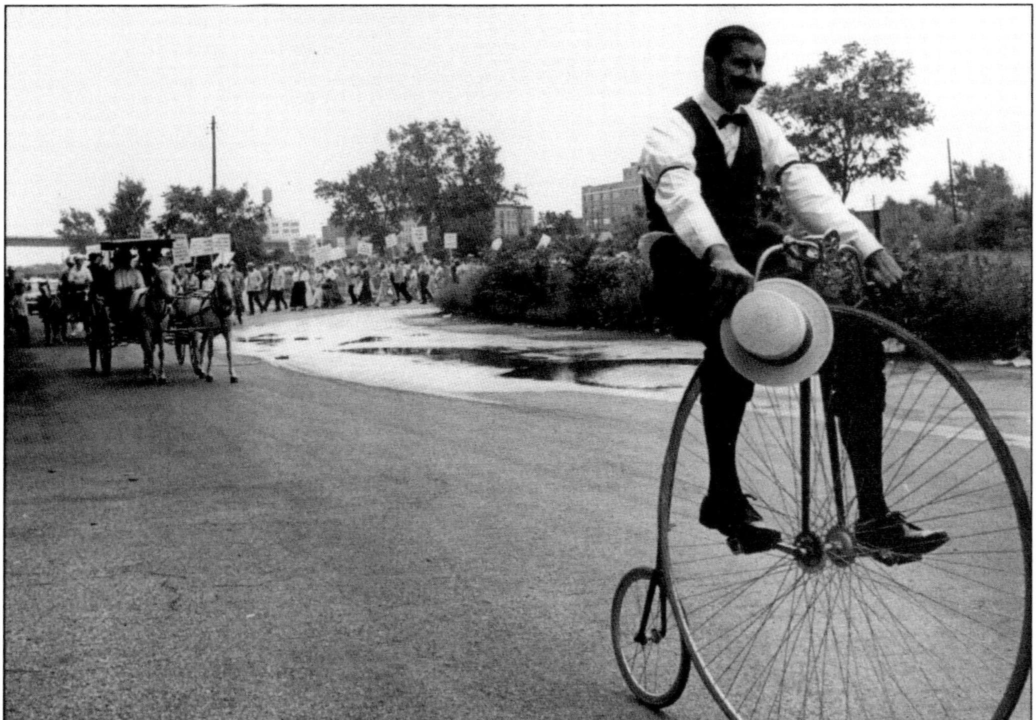

The parade marched through Toledo to the Ash Street plant of New England Glass. The parade route was three miles and lined with cheering Toledoeans. (Courtesy of the University of Toledo's Ward Canaday Center.)

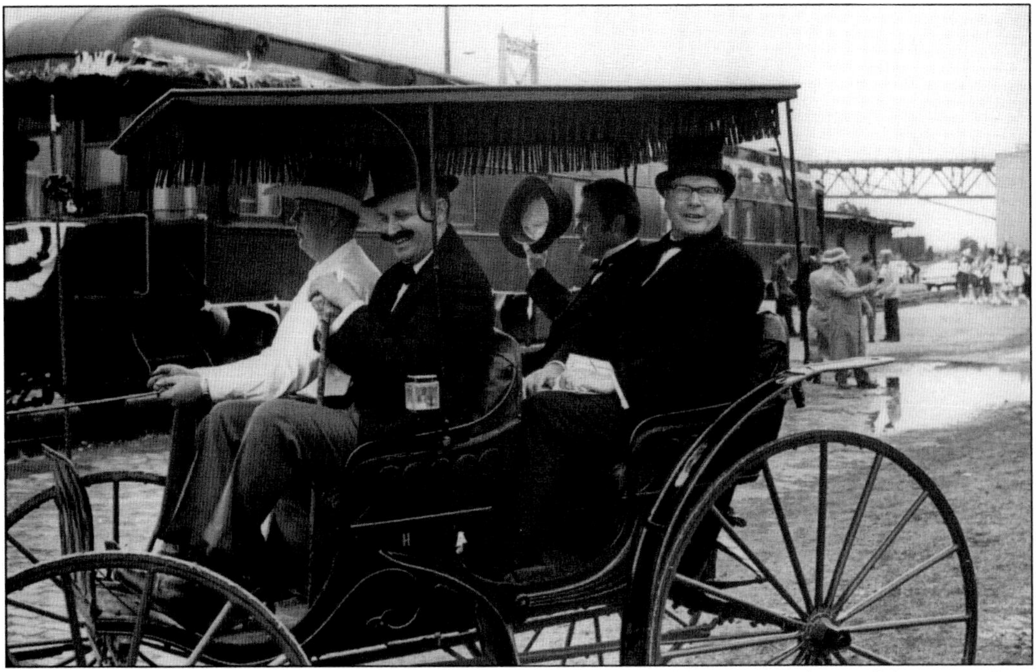

The train trip took 36 hours from Boston to Toledo in 1888. Libbey stepped out of the train first, followed by over 100 Irish glassworkers.

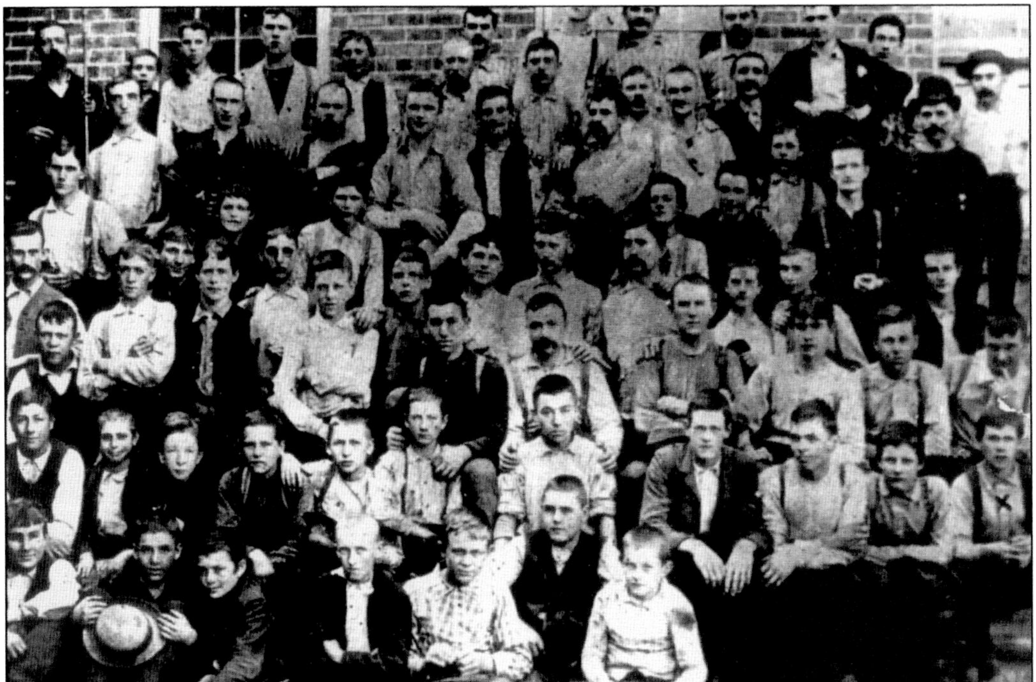

Here is an 1890 photograph of Libbey glassworks employees. Mike Owens is in the top row, second from right. Most of the original workforce of 1888 had returned to Boston due to homesickness or at the request of the new foreman Michael Owens. The original site included 50 lots for worker homes.

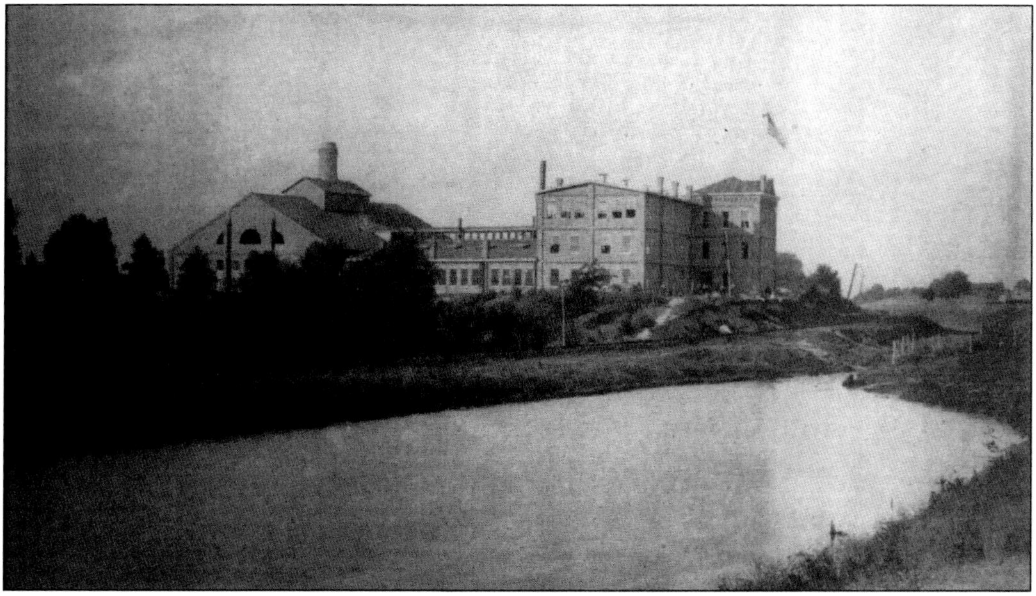

The plant opened in 1888, building had been delayed by the great blizzard of March 11, but breaking ground on St. Patrick's Day (March 17) was considered a lucky omen by the Irish workers from Boston. The plant produced its first melt in August 1888 and employed 250 employees. The site was originally a farm, which was being used as a baseball field.

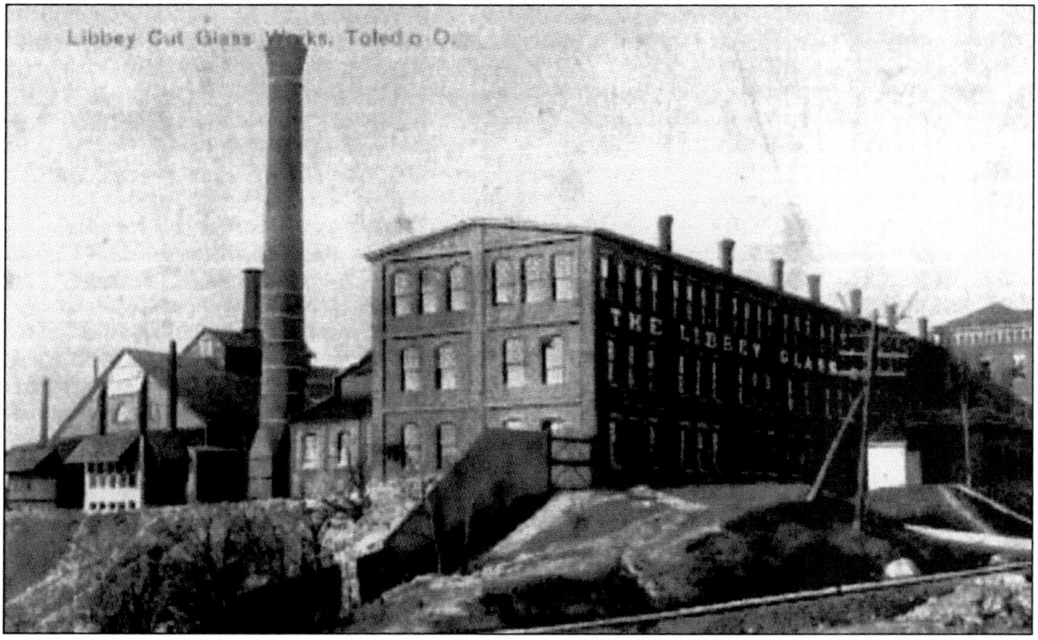

The original plant had a 13-pot furnace with each pot having a 2,500-pound capacity. The official 1888 name was the New England Glass Works of W. L. Libbey and Son Company; it was now known as Libbey Glass Company. The plant produced handblown flint glass with cut and engraved patterns. (Courtesy of the University of Toledo's Ward Canaday Center.)

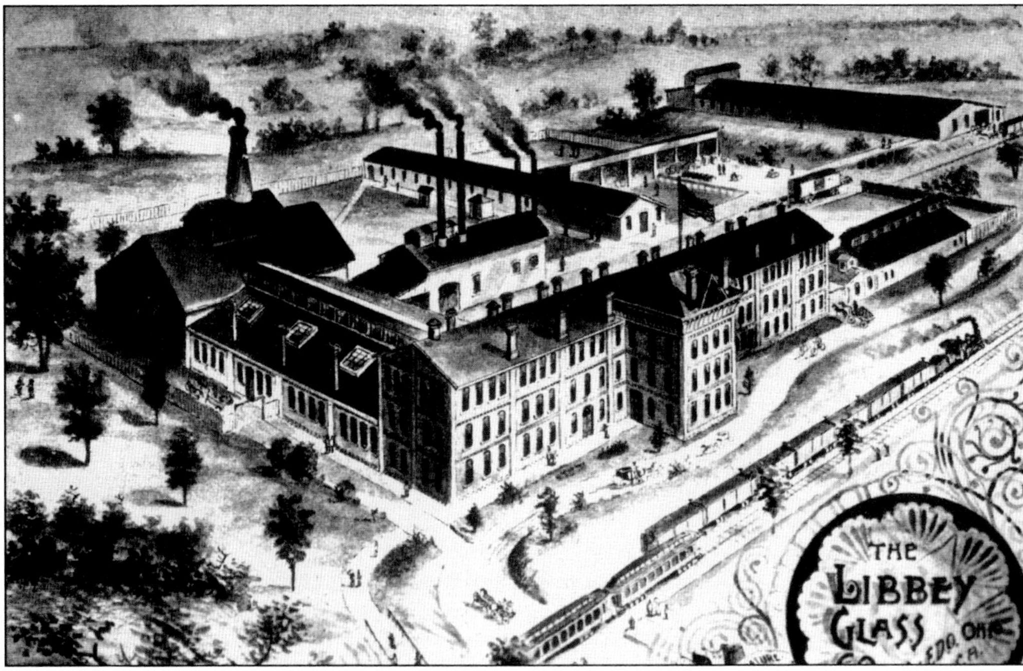

By the end of the century, Libbey Glass was one of the world's leading cut glass manufacturers. Some of Libbey's finest cut glass sold for $30, an amazingly high sum of money at the time. The plant also produced a small amount of handblown bottles and light bulbs as well. Libbey helped develop the surrounding neighborhood for his Irish workmen from Boston.

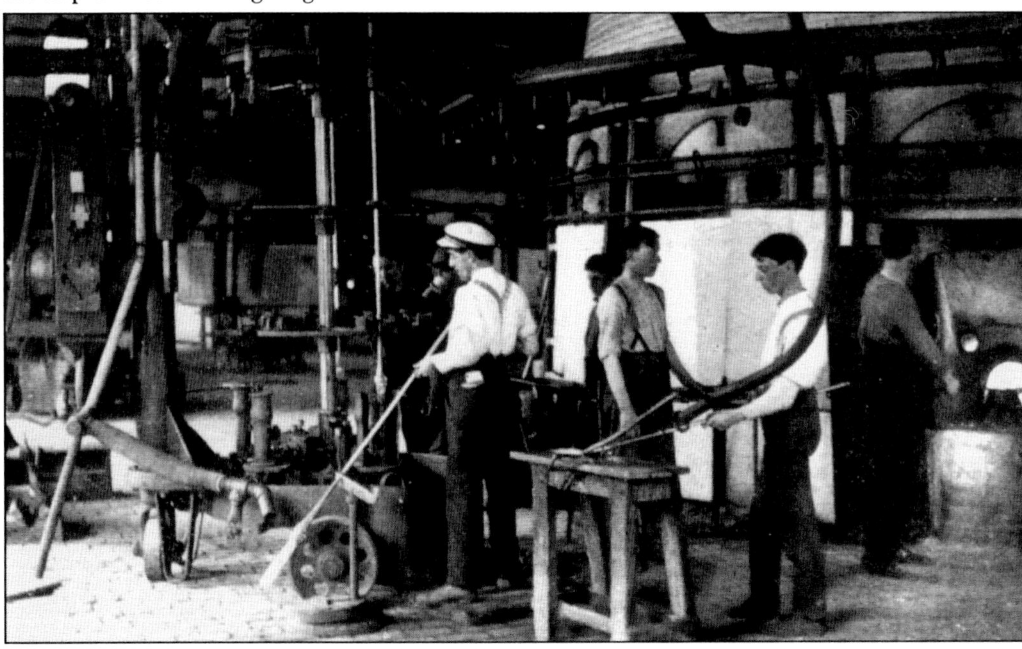

A gang, or glassblower's crew, worked one of 7 to 10 furnace openings. A gang consisted of five to seven workers. The master glassblower was in charge of the gang. The next most important member was the gatherer, who gathered molten glass from the end of a rod known as a pontil. It was then transferred to the master's blowpipe.

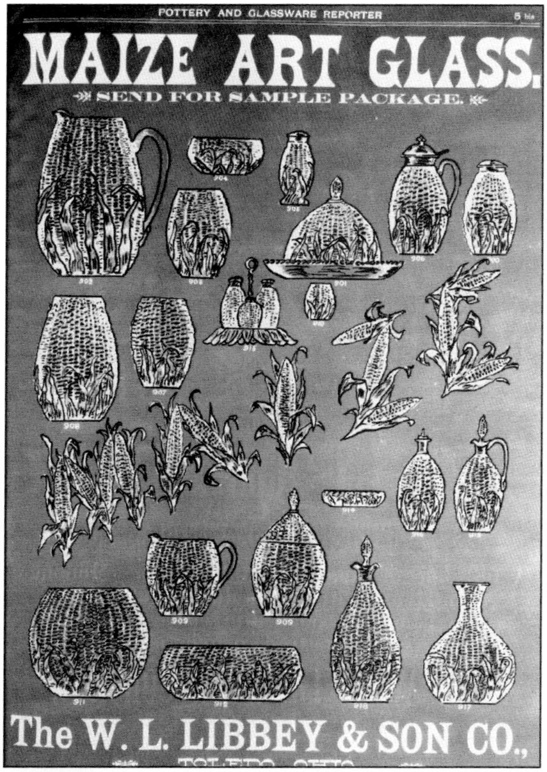

This 1890 advertisement illustrates Libbey's first Toledo pattern honoring Toledo's corn heritage. The company was originally known as W. L. Libbey and Son until 1892. The maize pattern would be one of Libbey's best sellers. Pieces were sold both as pressed glass and some cut pieces. (Courtesy of the University of Toledo's Ward Canaday Center.)

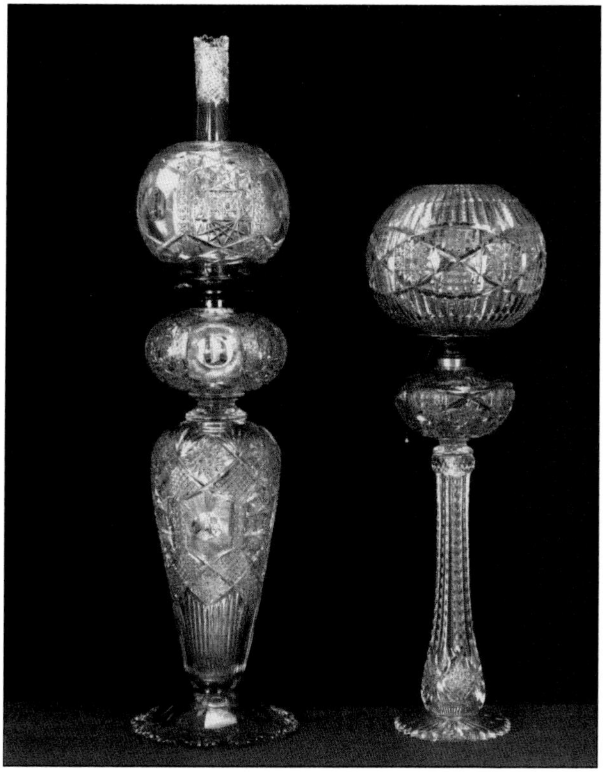

Two of the lamps produced by Libbey glassblower C. J. Nolan for the 1893 Chicago World Fair are seen in this image. These lamps became known as Nolan lamps. The lamp on the left was known as the sunburst pattern, while the one on the right standing 40 inches was the Ellsmere pattern, designed for the Chicago World Fair. (Courtesy of the Hancock Historical Society.)

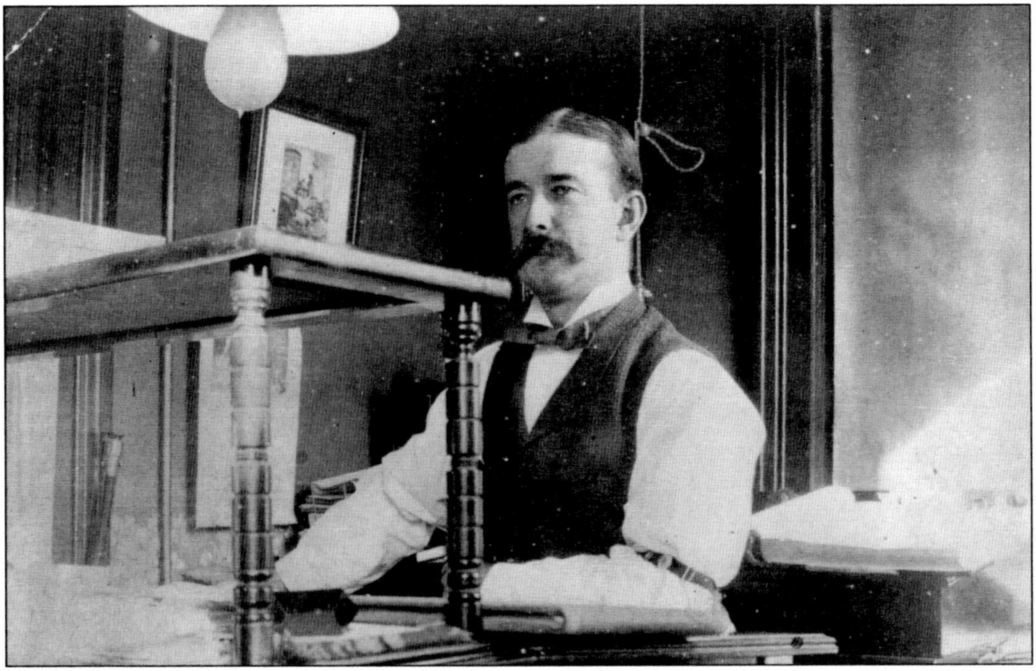

William Donovan was one of the key administrators that Libbey brought from New England Glass. For many years, Donovan was chief financial officer of Libbey Glass. He also was a director for Toledo Glass in 1895 and would ultimately rise to president of Libbey Glass in 1929.

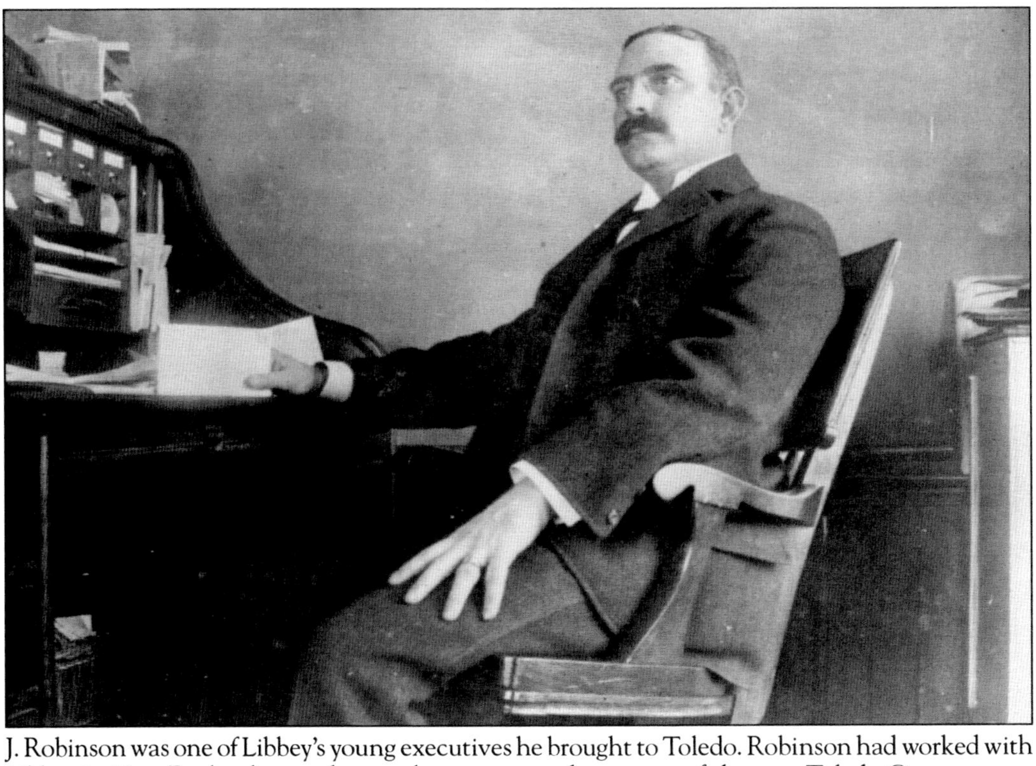

J. Robinson was one of Libbey's young executives he brought to Toledo. Robinson had worked with Libbey in New England as a salesman but was named treasurer of the new Toledo Company.

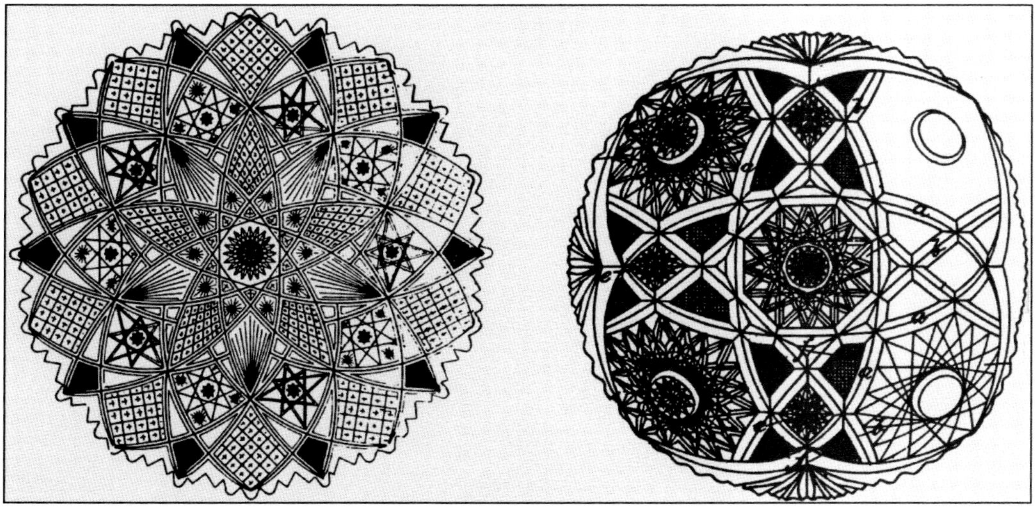

The chrysanthemum pattern shown above was by Libbey glassmaker William Marrett. Marrett came with Libbey to Toledo. The above pattern was patented in 1887. The pattern on the right was known as star diamond.

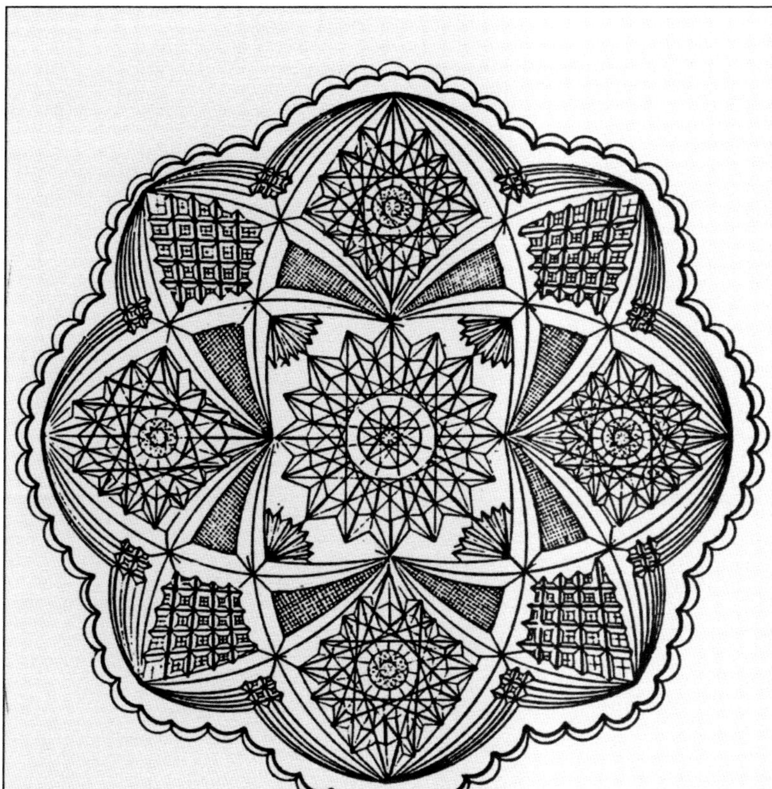

Another of Libbey's glasscutters, William Anderson patented the Corinthian pattern. Anderson arrived with Libbey in 1888 as superintendent of the cutting department.

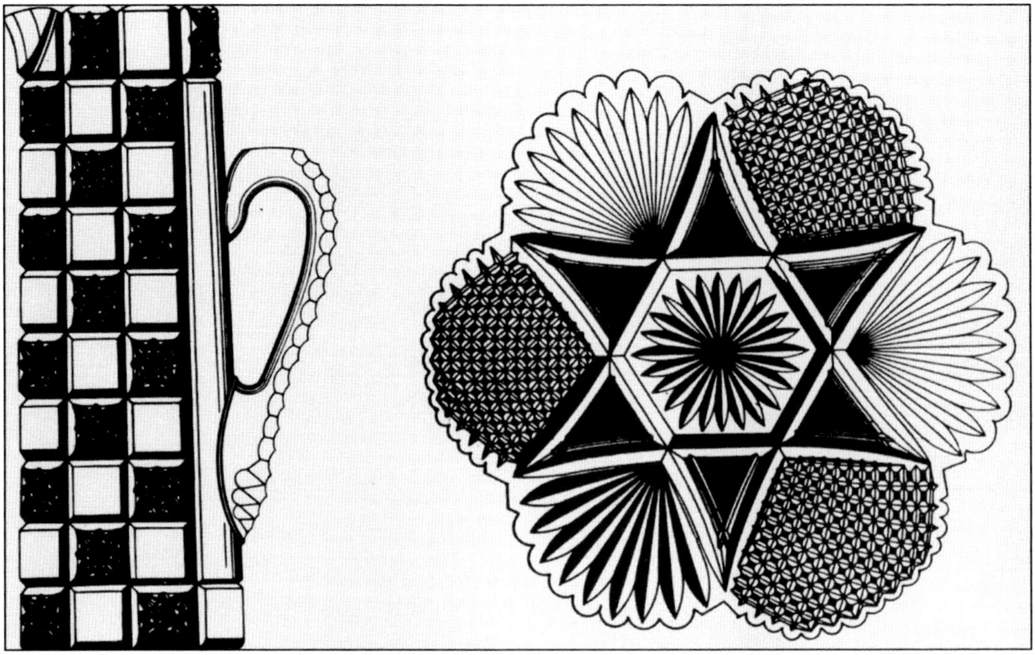

Sol Richardson had worked for Edward Libbey in the 1880s as a salesman. Richardson was not a glasscutter. He was one of Libbey's trusted executives who would become president of Libbey Glass. He patented the above pattern in 1889.

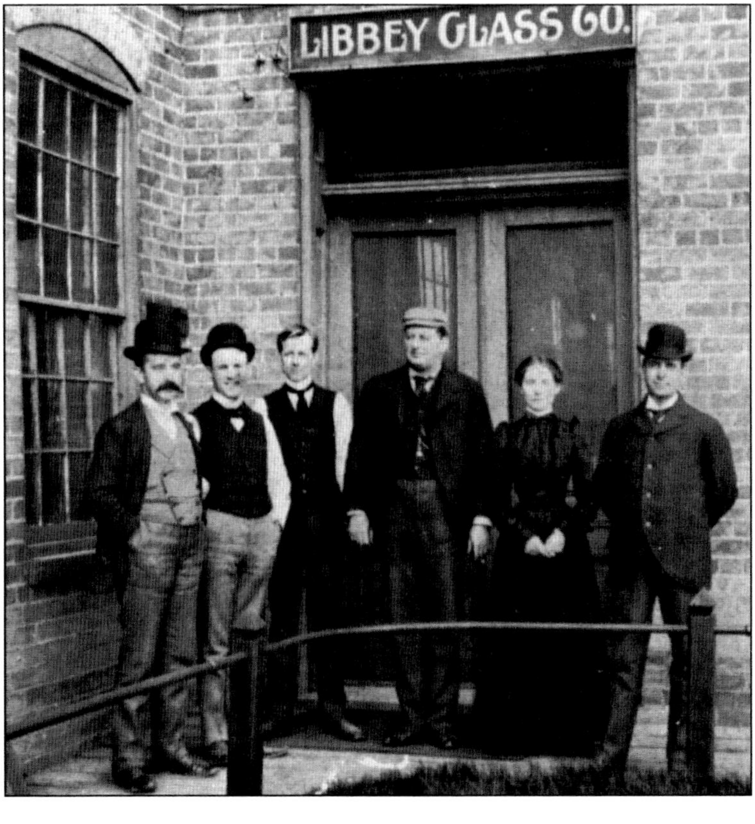

Seen here is the Libbey Glass company in 1892. Pictured from left to right are William Donovan (cutting department foreman), C. Robb (clerk), Charles Oswald (clerk), Sol Richardson (company secretary), Lillian Happin (stenographer), and J. Robinson (treasurer). (Courtesy of the Toledo Museum of Art.)

Four

Libbey Glass Prospers

The key to northwest Ohio becoming the glass capital was Libbey Glass. Libbey Glass would struggle for the first two years, until Libbey hired a young glassmaker from West Virginia's Hobbs-Brockunier. In a few short months, a young Michael Owens had turned the plant around, enabling Libbey to at least meet the payroll.

In late 1890, Corning Glass, the major supplier of light bulbs to Edison Electric, was embroiled in an "open shop" controversy with its union. A strike shutdown at Corning forced General Electric to search for a new supplier. Libbey, going back to its New England days, had supplied General Electric a small amount of bulbs. When General Electric contracted him for full production, the Toledo plant lacked capacity. The Corning union represented by the American Flint Glass Workers also asked Libbey for a deal where they would send Corning glassblowers to help until the strike was over. Libbey rented a plant in Findlay to manufacture bulbs. By 1892, at the end of the strike, the plant had returned a huge profit to Libbey, assuring Libbey Glass's survival. The Findlay operation was shut down, but Libbey would remain a suppler of bulbs for years to come.

Libbey, buoyed by his Findlay plant success, proposed to his board the plan to construct a large exhibition at the 1893 Chicago World's Fair. The board called the idea "foolish" and "a gamble." Libbey was forced to finance it himself with the help of Toledo banker John Ketcham. Libbey put his young manager, Michael Owens, in charge of building a complete glass plant on the fairgrounds. A huge 10-pot furnace was built in a "palatial palace." The huge plant was on the midway and could hold 2,000 visitors at a time. It was one of the most popular exhibits, with over 100 glass craftsmen being brought in from Toledo. Some of the best-selling pieces were cut patterns of Joseph Rosenberger, Libbey's glass artist. In the end, the fair was not only a financial success but also a huge marketing success, making the Libbey name known around the world.

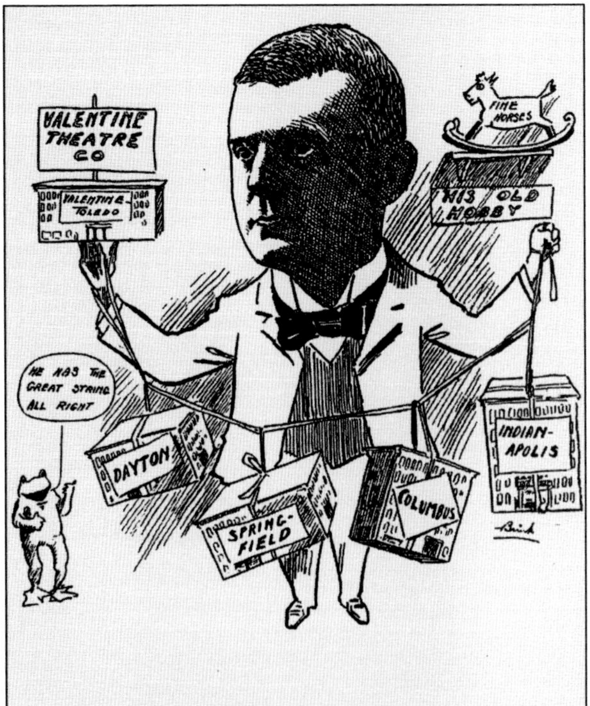

George Ketcham
Owner Valentine Building.

George Ketcham was the son of Valentine Ketcham, builder of the Valentine Theater. George was a key real estate investor and politician who supplied much of the land for Toledo's glass industry. It was a personal loan from Valentine to Libbey that allowed for the Libbey glass exhibition in the 1893 world's fair, which turned around the company. Without his support, Libbey Glass would have failed in 1894.

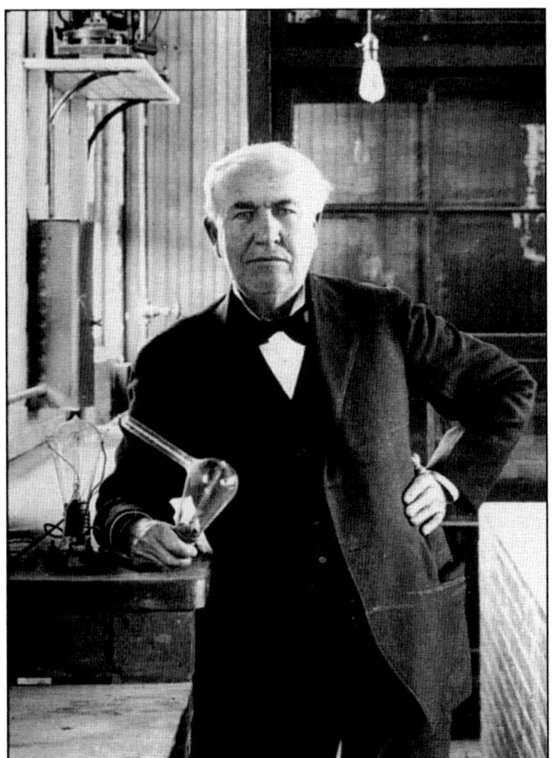

Originally, in the 1880s, Thomas Edison had laboratory glassblowers making bulbs from glass tubing (note the shape of the bulb in Edison's hand). New England had established a working relationship with Edison's Menlo Park in the early 1880s. (Courtesy of the Henry Ford Museum.)

A strike at the Corning Glass light bulb plant in New York resulted in a transfer of business to a Libbey Glass plant in Findlay in 1891. Known as the Richardson Plant, it was managed by Mike Owens. The American Flint Glass Workers Union supported the transfer and supplied Corning glassworkers to the Finley plant. The booming business for light bulbs required the company to forgo the normal summer shutdown, giving workers a shorter vacation. A total of 19 workers died in a train wreck in Ravenna on their way to families in Corning. The union erected a monument in St. Mary's Cemetery in Corning, New York.

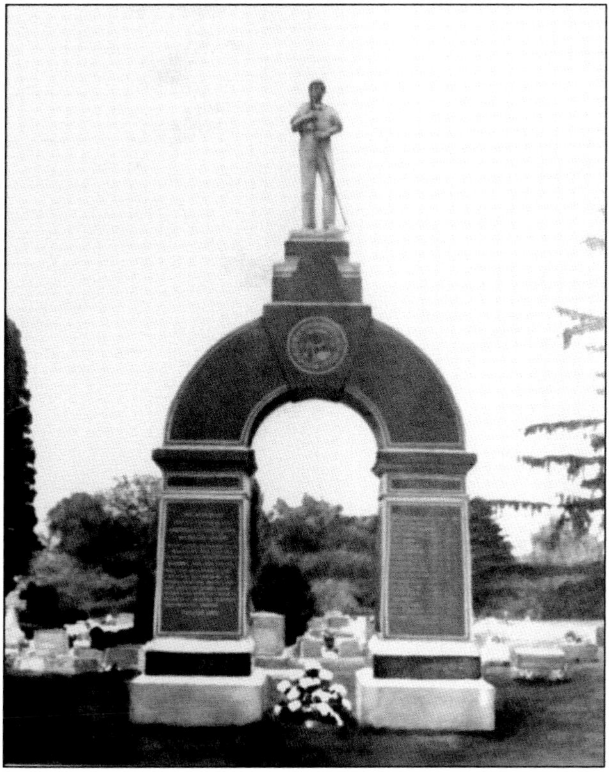

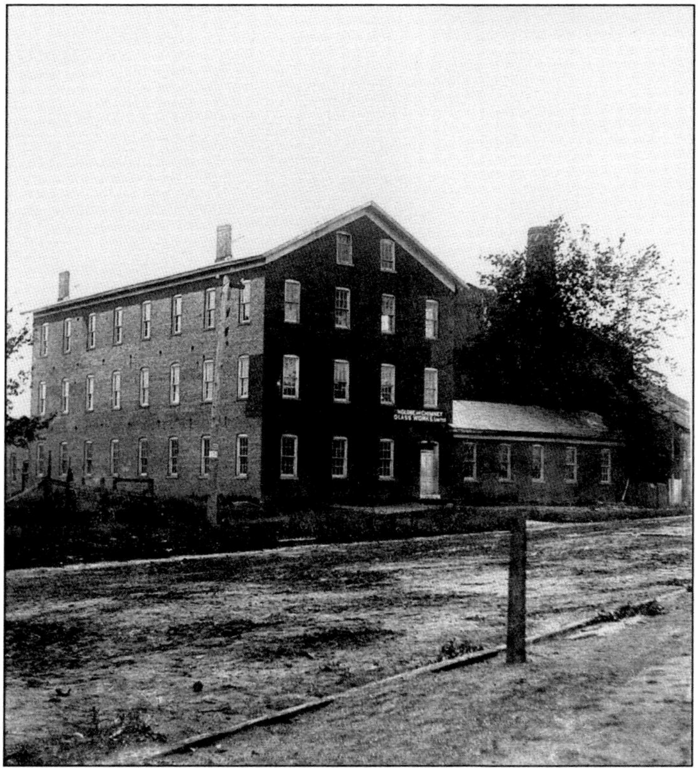

This well-built glassworks started production of oil lamp chimneys and some electric light bulbs in 1889, after the company's glassworks in Ravenna burned down. Most of the 200 workers were from Ravenna and returned when the glassworks was rebuilt. This East Main Street factory would become famous when Libbey rented it to produce electric light bulbs during the Corning Glass strike in 1891. (Courtesy of the Hancock Historical Society.)

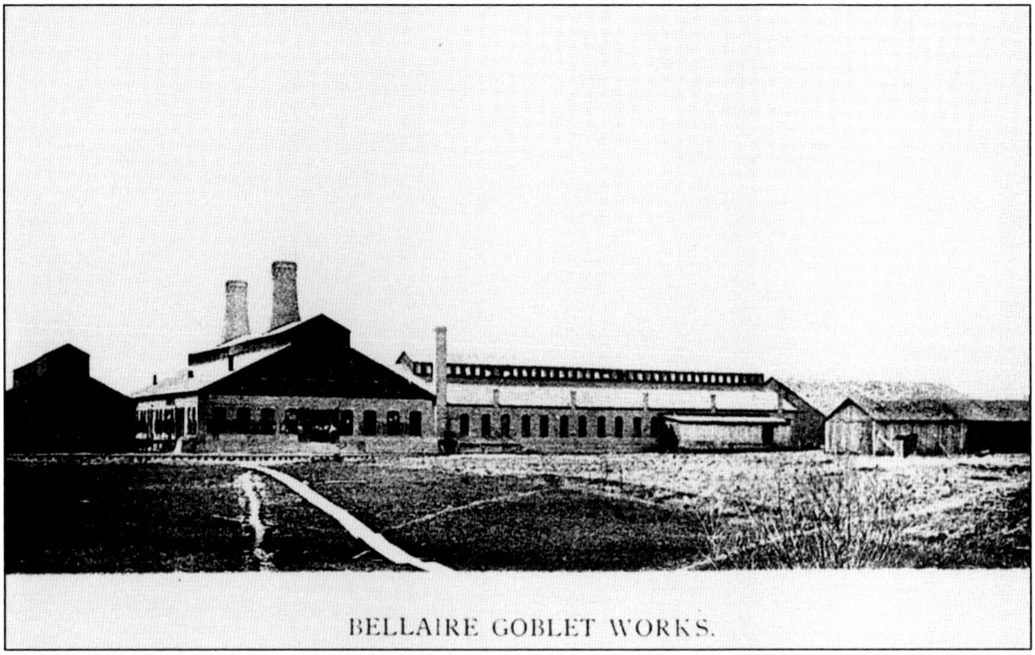

In 1888, this 30-pot furnace factory, which was the largest in the world, came to Findlay. Built on College and Davis Streets, the plant operated as a premier tableware producer. In 1891, it became Factory M of the United States Glass merger.

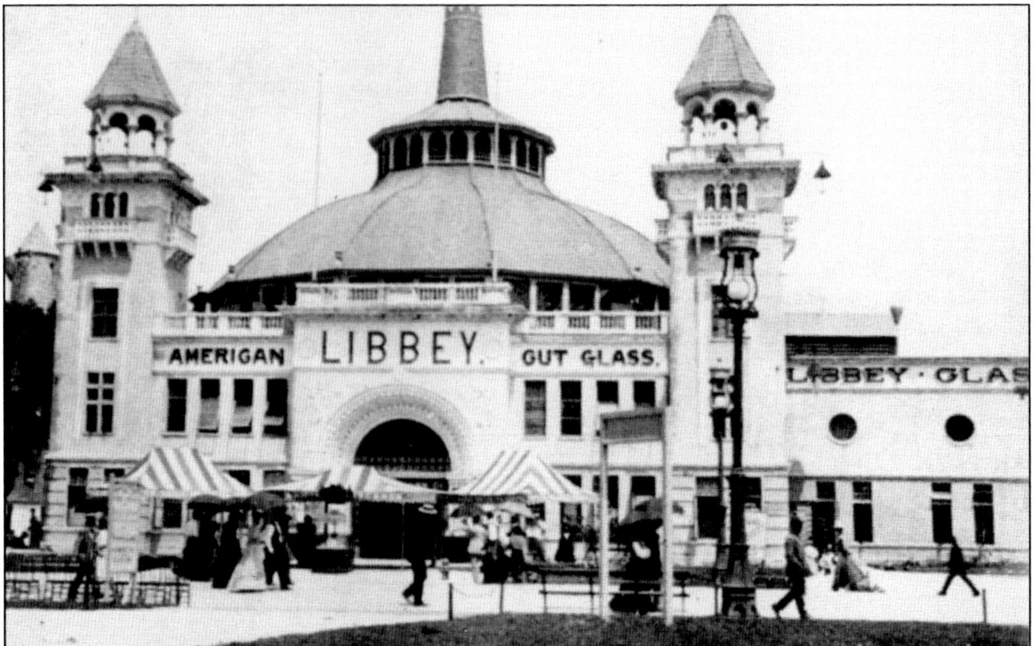

Libbey found no support for his world fair exhibition from his board of directors. He had to secure a personal loan from George Ketcham to build it. The building and management of this complete glassworks rested on the shoulders of Libbey's youngest manager—Mike Owens. (Courtesy of the University of Toledo's Ward Canaday Center.)

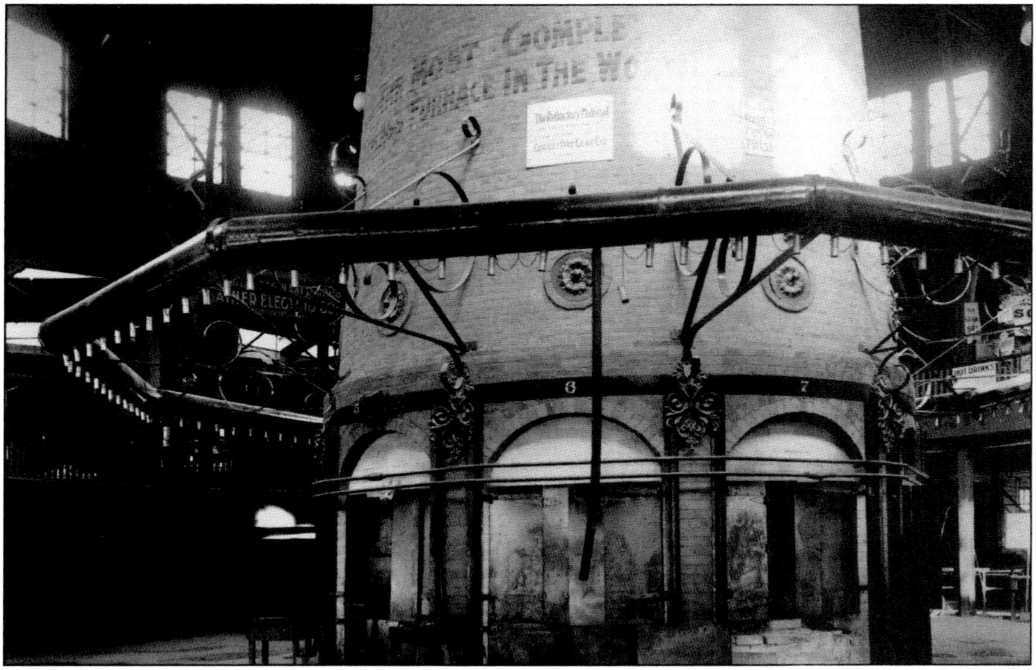

The exhibition was a complete glassworks. Libbey brought 40 glassblowers, including John Rosenberger, one of the world's best, to operate the 10-pot furnace. In addition, Libbey brought his best glasscutters. The foolish gamble of Libbey and Owens would pay dividends for years to come.

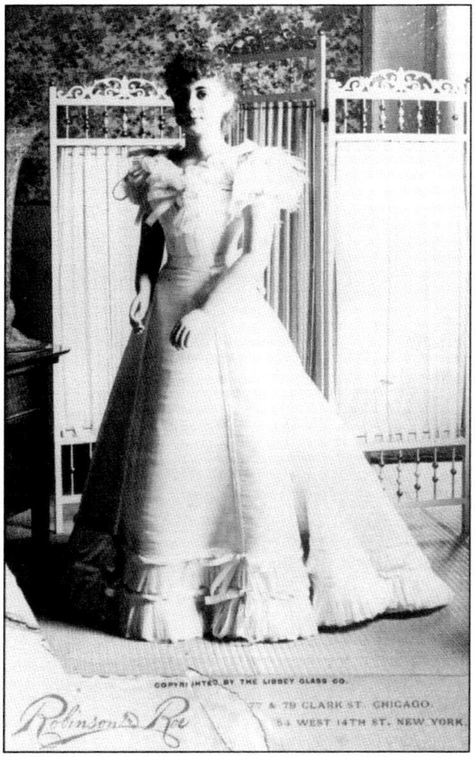

Early on at the fair, the Spanish princess visited and commissioned a glass fiber dress like the one displayed. Libbey was fascinated by glass fibers and started a secret operation of German glass artists to develop it further. The operation failed after a few years, but in the 1940s, Libbey Glass would develop a successful process. (Courtesy of the Toledo Lucas County Library.)

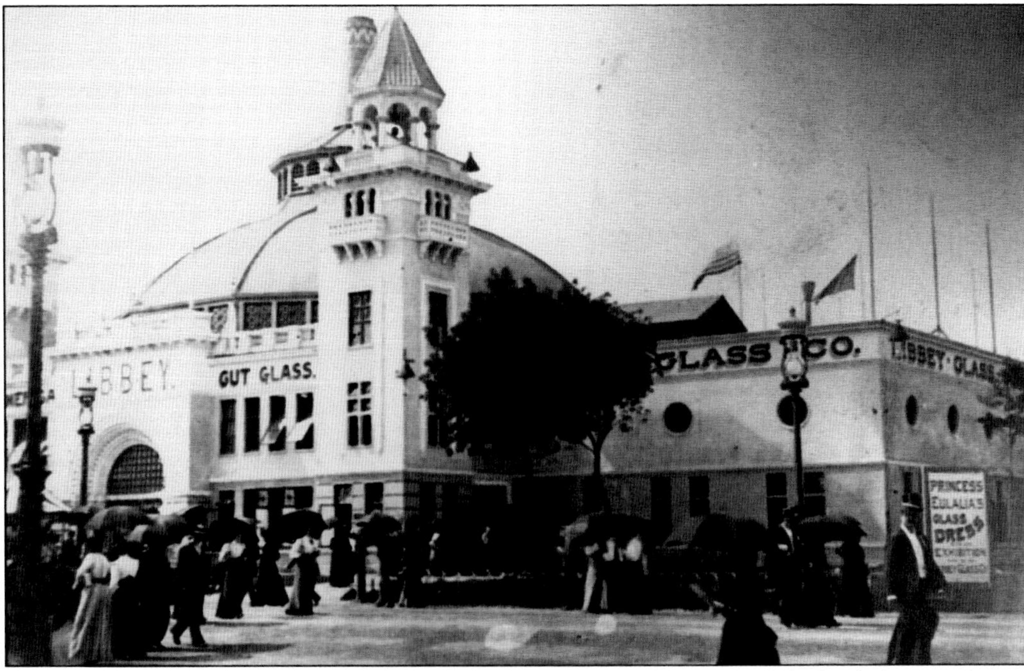

Note the billboard advertisement of the Princess Eulalie dress, which helped the number of visitors to skyrocket after a slow start. The top story of the pavilion was actually a suite of rooms for the Libbeys to live in. (Courtesy of the Toledo Museum of Art.)

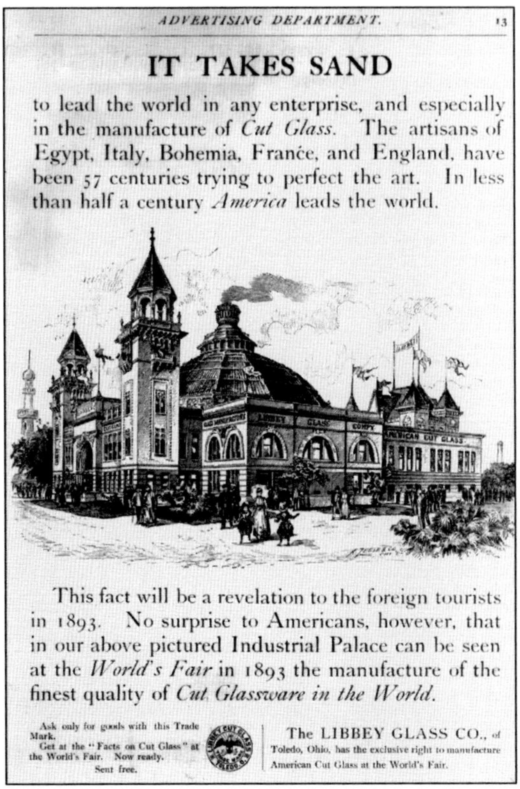

ADVERTISING DEPARTMENT. 13

IT TAKES SAND

to lead the world in any enterprise, and especially in the manufacture of *Cut Glass.* The artisans of Egypt, Italy, Bohemia, France, and England, have been 57 centuries trying to perfect the art. In less than half a century *America* leads the world.

This fact will be a revelation to the foreign tourists in 1893. No surprise to Americans, however, that in our above pictured Industrial Palace can be seen at the *World's Fair* in 1893 the manufacture of the finest quality of *Cut Glassware in the World.*

Ask only for goods with this Trade Mark.
Get at the "Facts on Cut Glass" at the World's Fair. Now ready. Sent free.

The LIBBEY GLASS CO., of Toledo, Ohio, has the exclusive right to manufacture American Cut Glass at the World's Fair.

New England Glass had used the high-quality sand of the Berkshire deposit in New England. Libbey continued to ship this sand to Toledo, but mixed it with the high-quality sand of Holland and Monclova deposits near Toledo. The real secret of Libbey Glass's brilliance was his extra shot of lead oxide (24 to 30 percent).

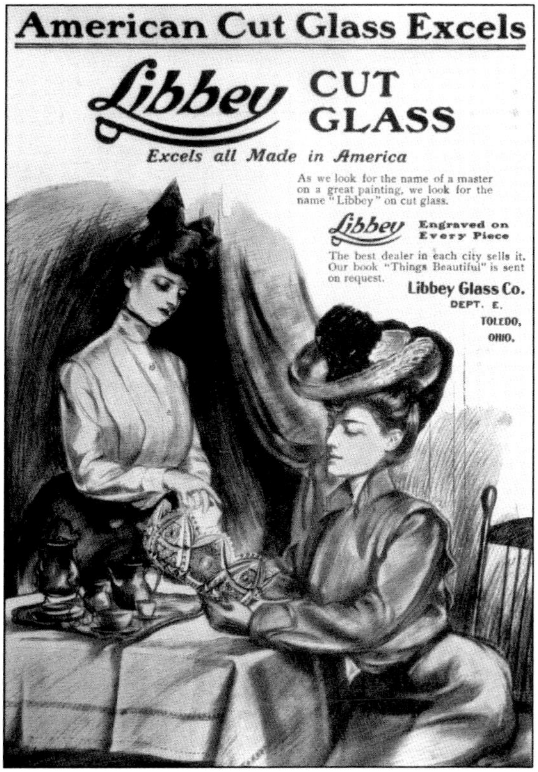

Note the change in the Libbey trademark to a sword. Throughout the 1890s and 1900s, Libbey Glass dominated the high end of the cut glass market in America. (Courtesy of the University of Toledo's Ward Canaday Center.)

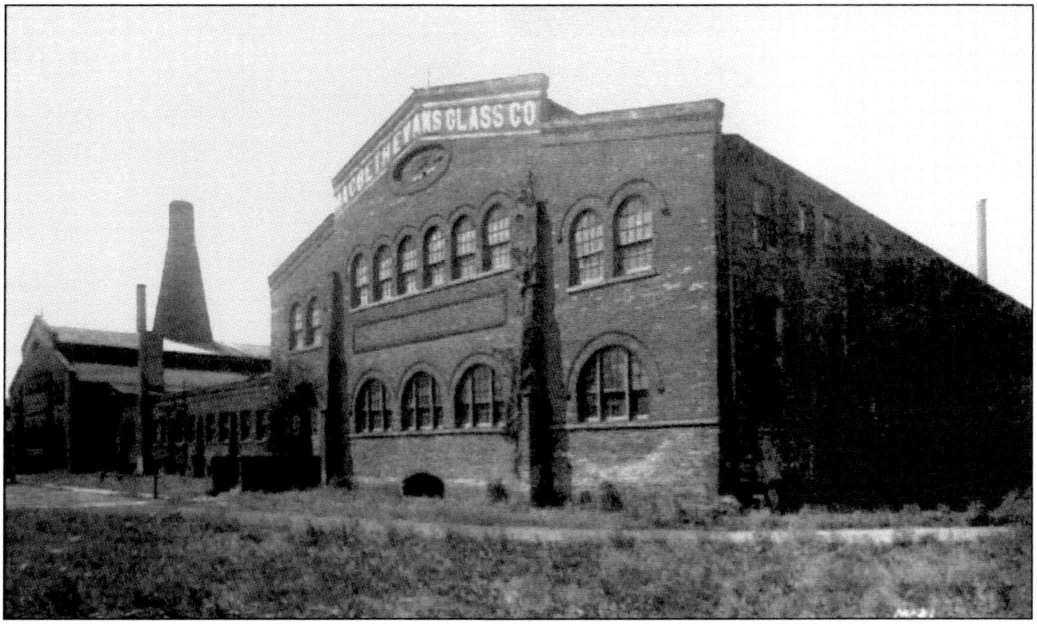

The Macbeth-Evans Company was actually owned by Libbey, Walbridge, and Owens (Toledo Glass Company). The plant had originally been built a few years earlier as the Illuminating Glass Company, which produced glass chimneys for oil lamps and light bulbs. Considered at the time the world's largest glass chimney factory utilizing semiautomated Owens glass machines. Mike Owens was named superintendent of the plant in 1899. The plant ran until 1925.

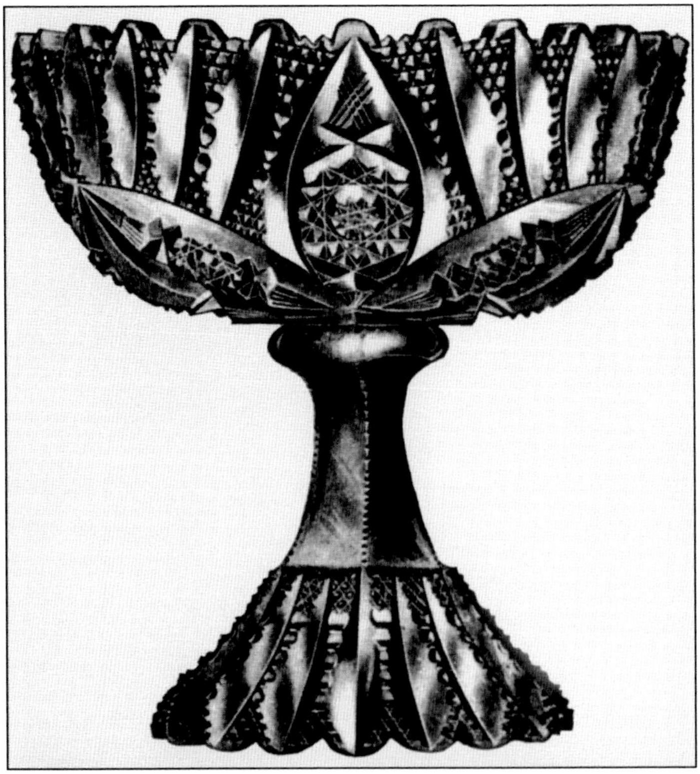

Libbey continued its dominance in tableware and cut glass through 1920.

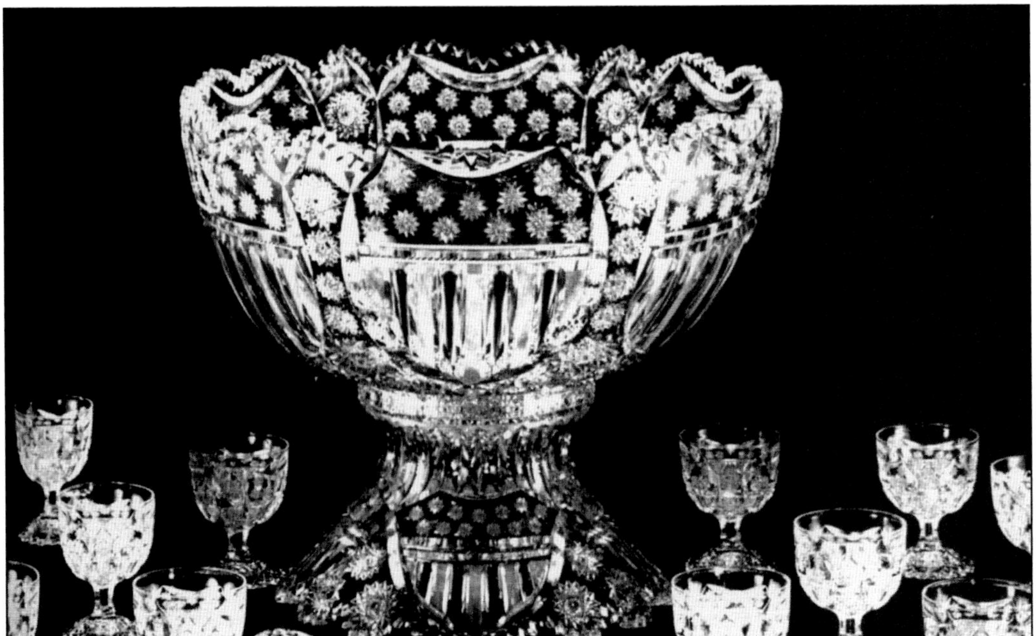

Libbey commonly made commemorative punch bowls. This one was made for President McKinley in 1898. It was the McKinley tariff that the rooting and growth of the American glass industry is attributed to. The bowl was rough abrasive wheel cut by Pat Walker and copper wheel finished by master cutter William Marrett.

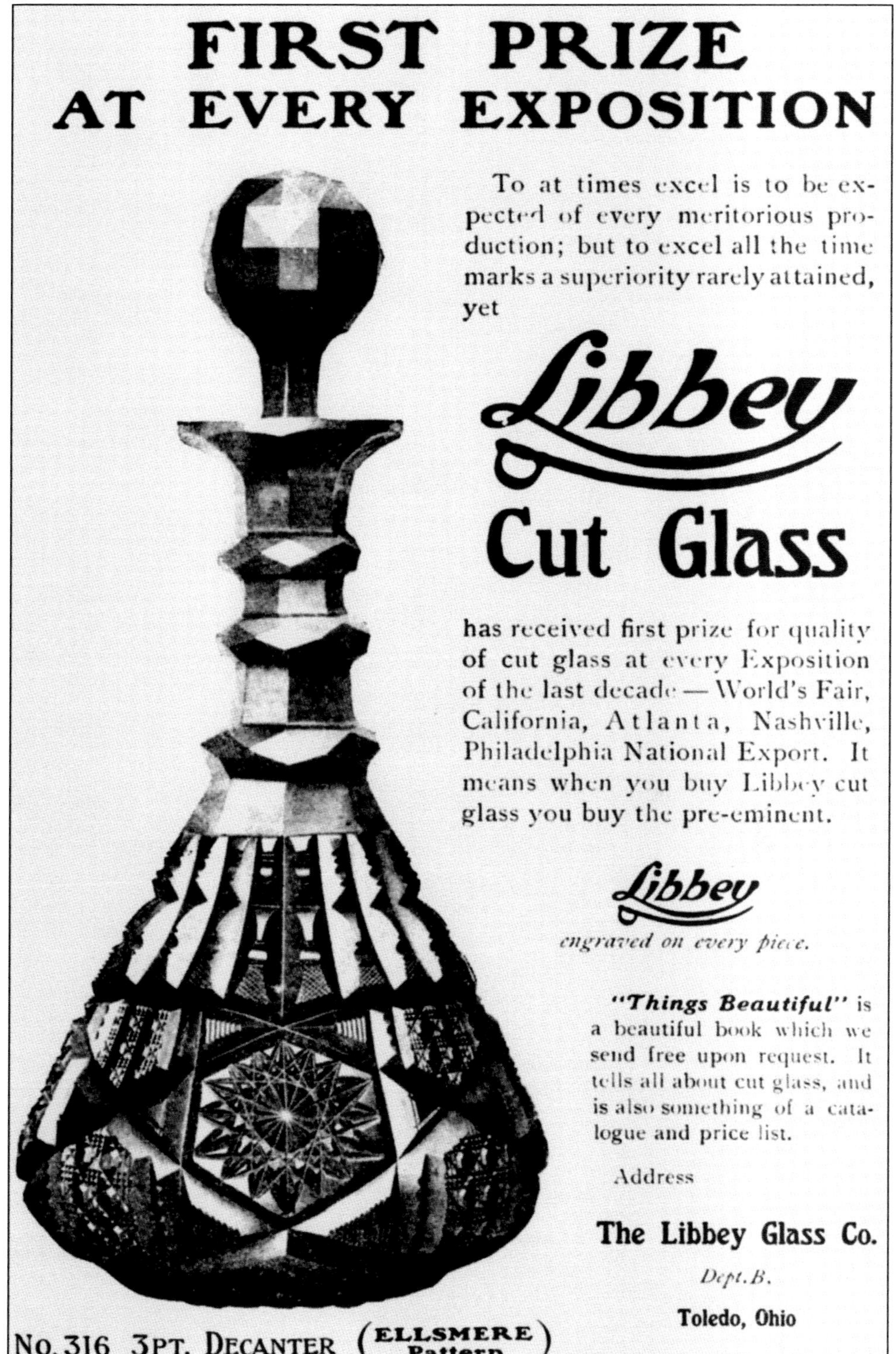

Edward Libbey was an aggressive user of advertising since his success at the 1893 world's fair. The Ellsmere pattern was also called the strawberry diamond.

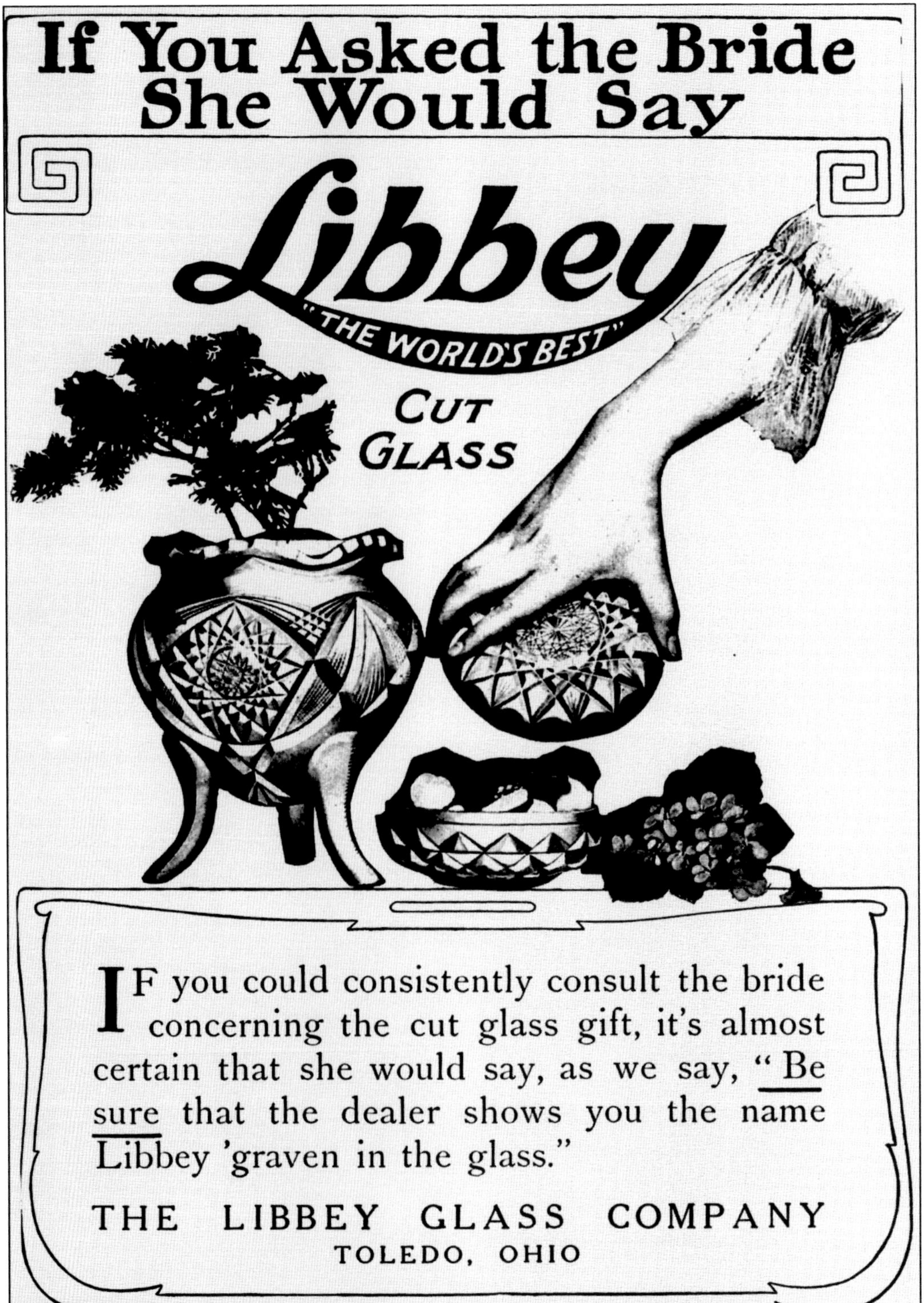

Wedding gifts had been a market segment for Libbey going back to New England Glass.

Five

Northwest Ohio Loses Natural Gas

A few months before Libbey Glass produced its first melt, Dalzell glass had established itself as a premiere producer of blown and pressed glass. Dalzell's 1888 glass plant was hailed as one of the country's largest, employing over 300 men and boys. Another prestigious company, Fostoria Glass was opened in Fostoria, Ohio. Similarly, Tiffin's Belgian Glass opened in July 1888. Belgian Glass produced cast colored cathedral glass and was the only glasshouse of its kind in the United States. Even Maumee Glass produced cylinder window glass before Libbey arrived. Most of the 30 some companies that beat Libbey to the area failed within a few years as the natural gas disappeared.

Another part of the gas boom and glass rush was the development of Baltimore and Ohio railroad stops in the Trenton gas formation. Bloomdale was one of these tiny towns. Welker, a junction of the Toledo and Ohio Central and the Baltimore and Ohio Railroads in southern Hancock County, also rose into a tiny mecca with several glass companies. North Baltimore was a junction of the Toledo, Columbus and Southern Railroad with the Baltimore and Ohio Railroad. North Baltimore attracted a bottle, window, and tumbler company bringing in about 500 workers.

In 1891, the gas pressure was failing throughout northwest Ohio and companies were failing as well. Another part of the problem was an overexpansion of the American glass industry in the Ohio-Indiana boom. David Jenkins, president of Findlay's first glass company—Columbia Glass, wanted to merge a group of companies to gain economies of scale. In July 1891, Jenkins merged 17 companies in Pittsburgh, West Virginia, and Ohio into United States Glass Company. Columbia Glass became United States Glass Factory J, Belaire Goblet Company in Findlay became Factory M, Nickel Plate Glass in Fostoria became Factory N, A. J. Beatty in Tiffin became Factory R, and Novelty Glass in Fostoria became Factory T. United States Glass was headquartered in Pittsburgh, and it never fully realized the benefits of the merger.

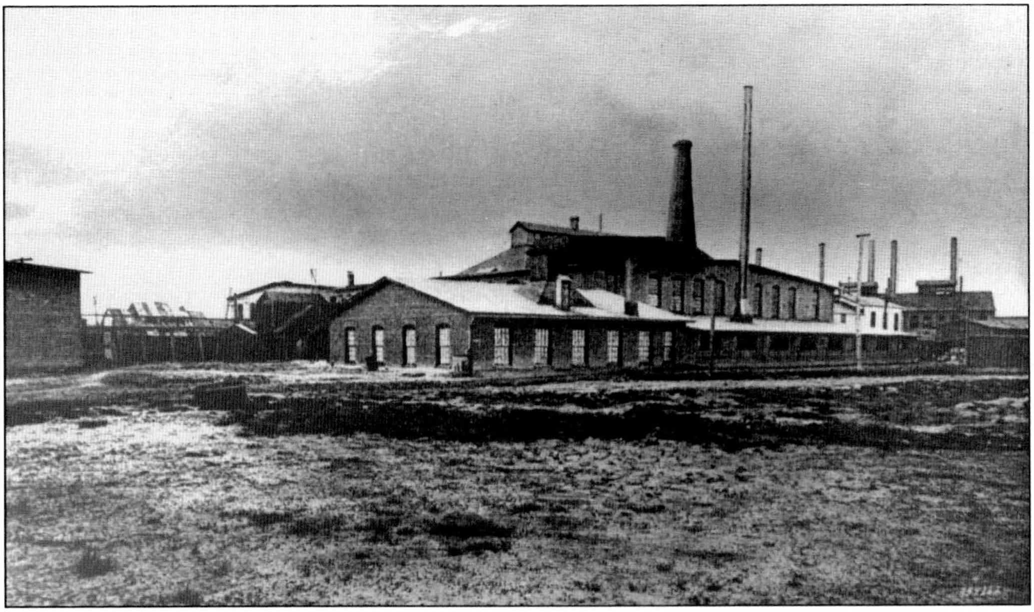

Originally opened as Fostoria Bulb and Bottle Company in 1900, Fostoria Specialty was actually owned by Cleveland's National Electric Lamp Company, which took over the Loudon plant in 1902. The plant was a major producer of bulbs and glass tubing for General Electric, until General Electric transferred the business to Libbey Glass in 1910. The Loudon plant then produced specialty bulbs for General Electric until 1920. The plant remained one of the few handblown bulb plants after the Libbey Glass automated machines took over the market in 1910.

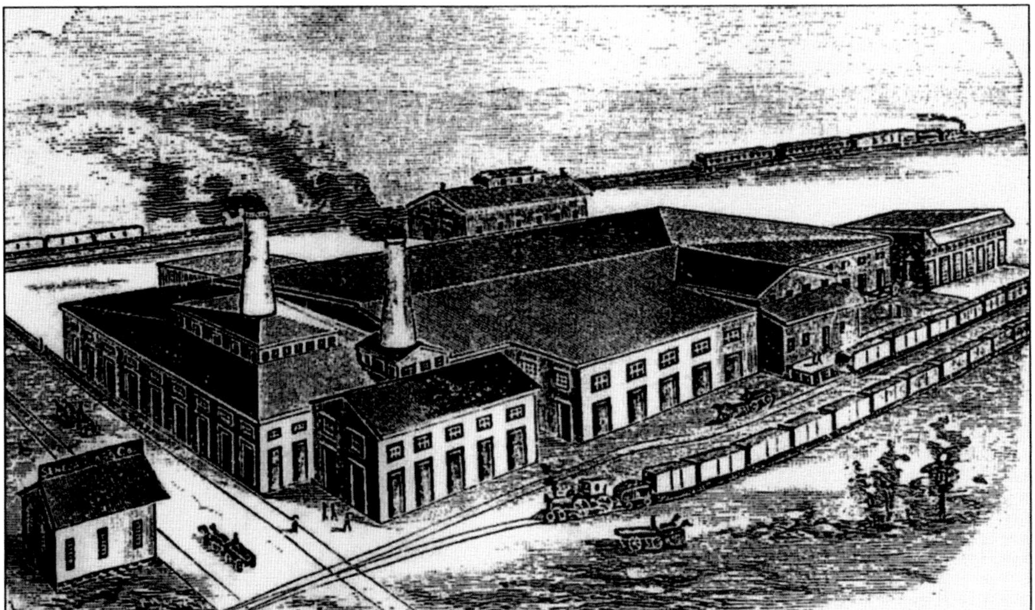

When the now famous Fostoria Glass Company left Fostoria after three years of operation in 1891, a group of highly trained immigrant German glassmakers from Cumberland, Maryland, took over the plant. Seneca glass at South Vine Street closed in 1905 and was taken over by Seneca Wire and Manufacturing Company. (Courtesy of the Kaubisch Memorial Public Library.)

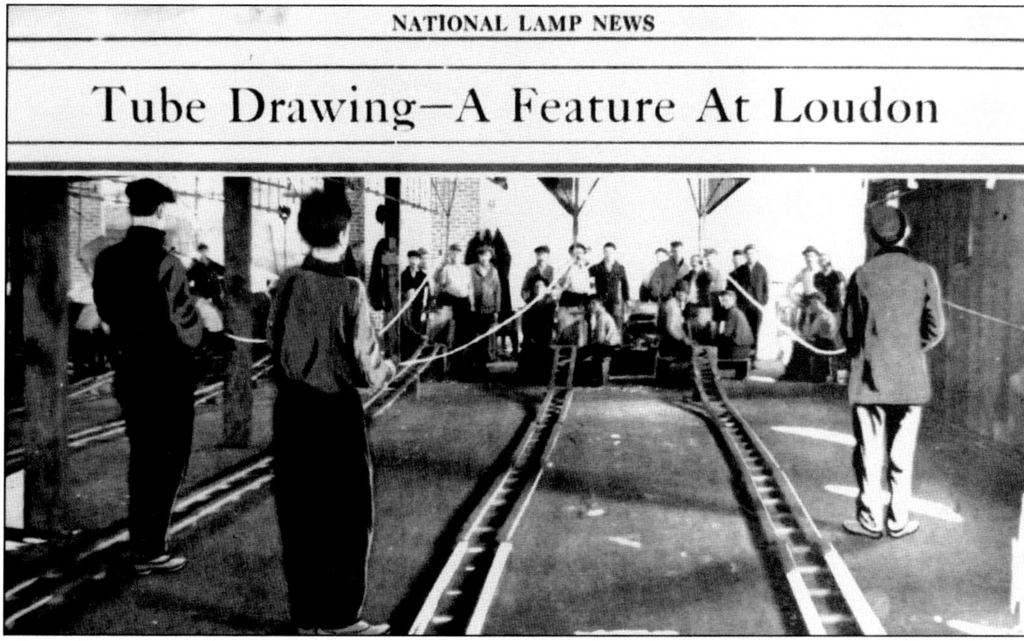

These boys are "walking" or pulling a 15-pound hollow ball of glass down a 150-foot alley. The hollow tube was maintained by a glassblower at the one end. In 1912, the Loudon plant produced 6,000 to 7,000 pounds of tubing a day. (Courtesy of the Kaubisch Memorial Public Library.)

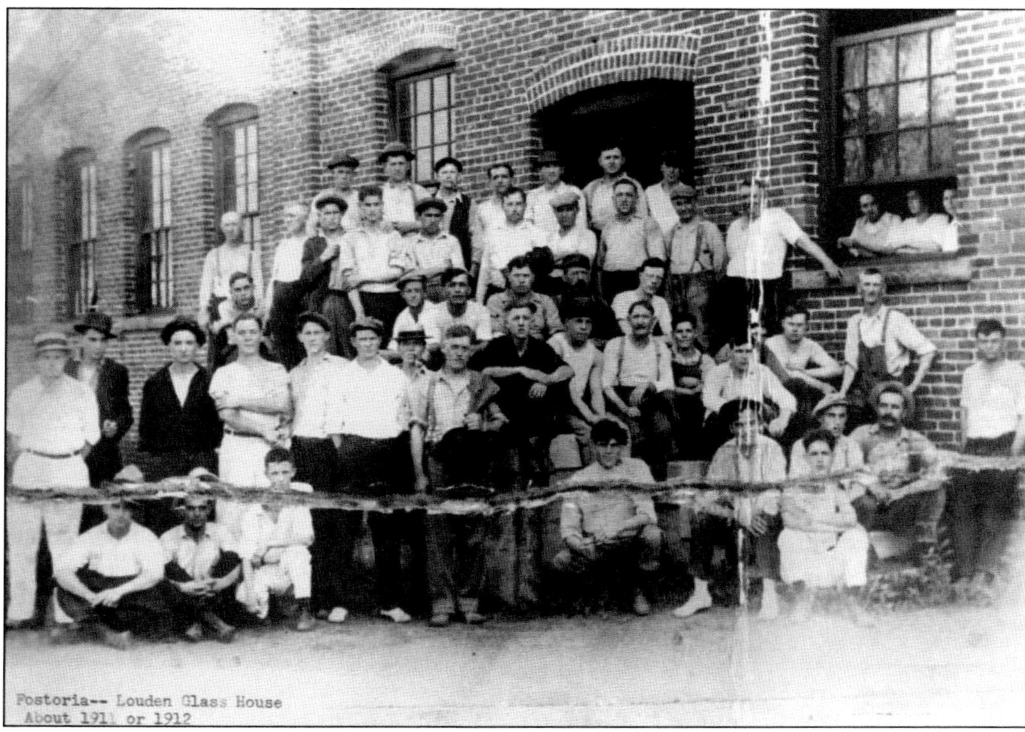

In 1911, the Loudon plant became part of General Electric. Employment dropped from several hundred employees to around 50 as tubing was eliminated and bulb production reduced. (Courtesy of the Kaubisch Memorial Public Library.)

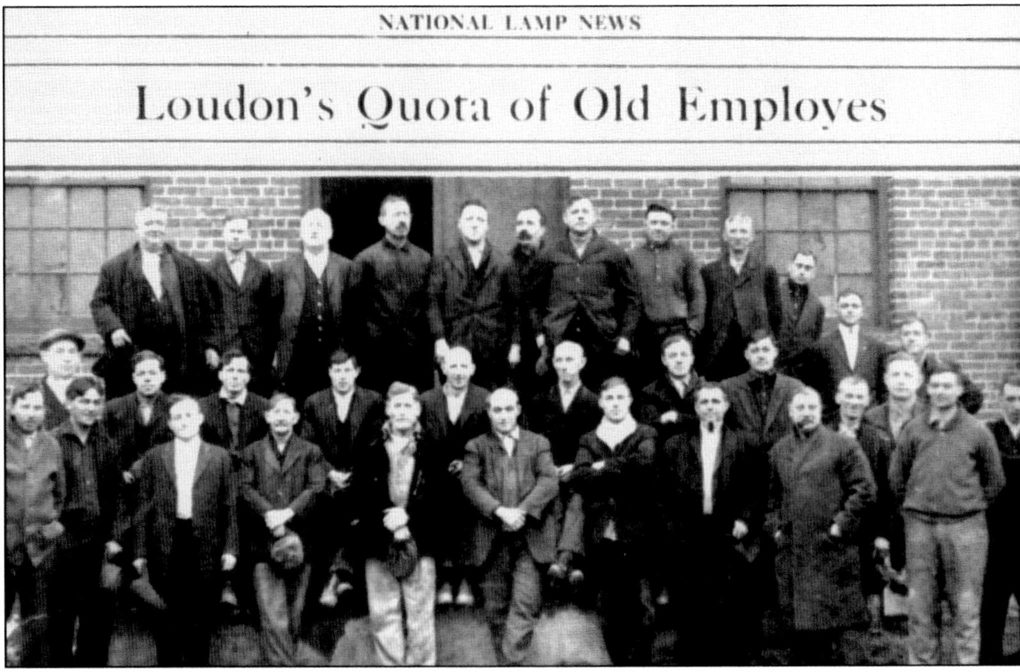

Glassmaking was a young man's career due to the injury rate, heat, and hard work required. The average career was about six years prior to 1900. Serious injuries were common, but it was the backbreaking hot work that shortened the glassworkers' productive years. The heat was so bad that the glass industry was the only industry to shut down in the summer.

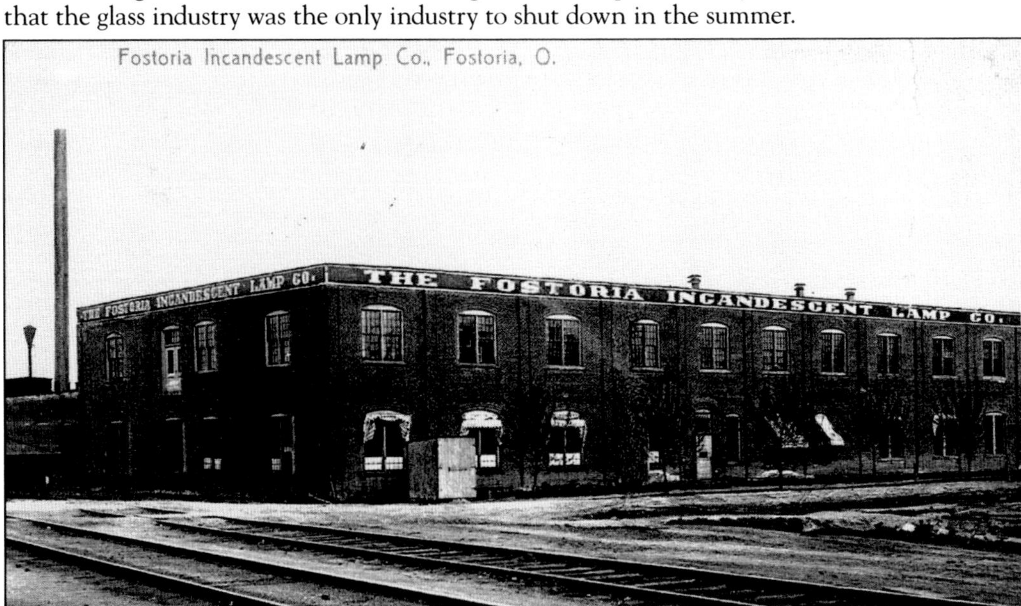

The plant was not actually a glasshouse, but a manufacturer of electric bulbs. Glass bulbs were shipped in barrels from Libbey Glass in Toledo originally, and later from the Loudon glasshouse plant. The plant consisted of an all-women assembly room, which accounted for most of the 120 workers, and an all-male vacuum room where the final lamp bulb was assembled. (Courtesy of the Kaubisch Memorial Public Library.)

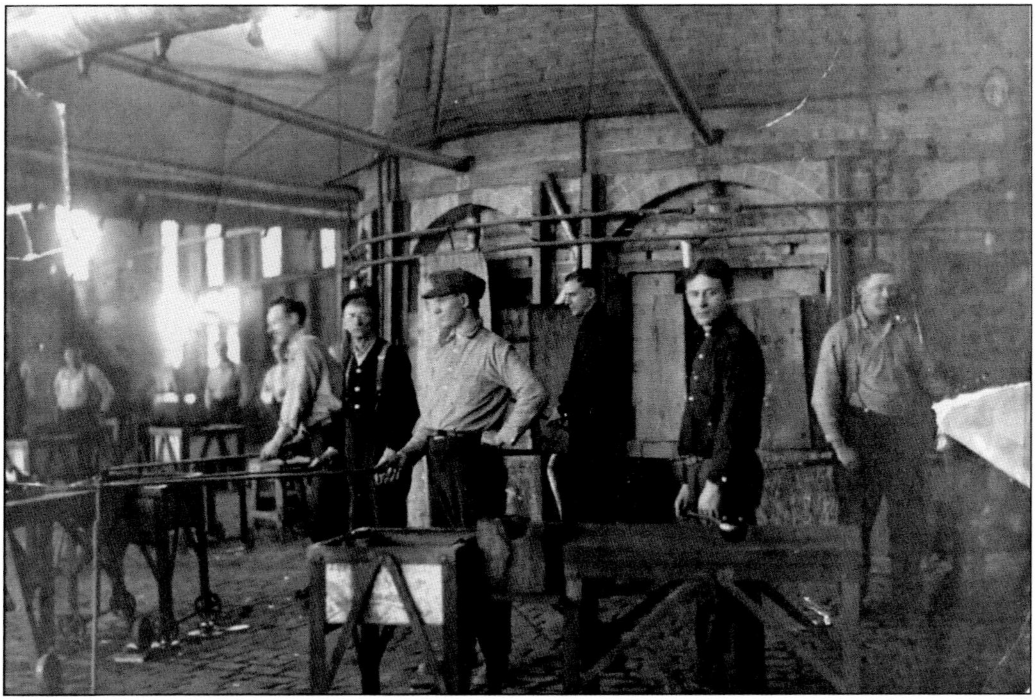

Fostoria Glass was the most famous of the Fostoria glass companies. The plant came to northwest Ohio in 1888 but returned to Moundville, West Virginia, in 1891 as the cheap gas supplies disappeared.

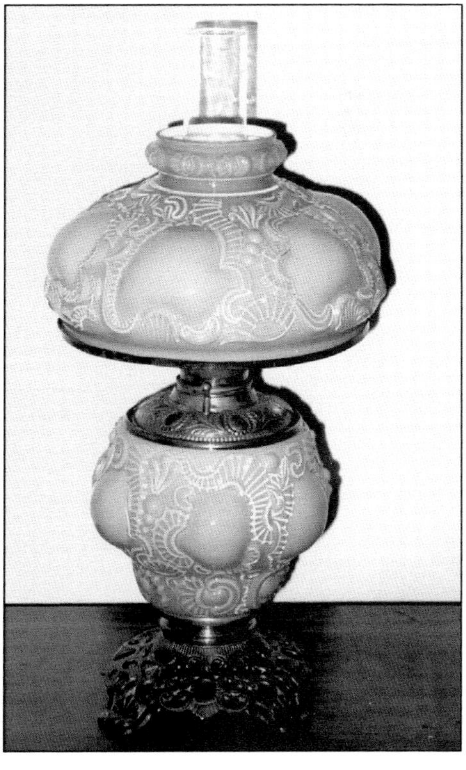

The custer pink lamp was typical of Fostoria Glass and Fostoria Shade and Lamp in the 1890s.

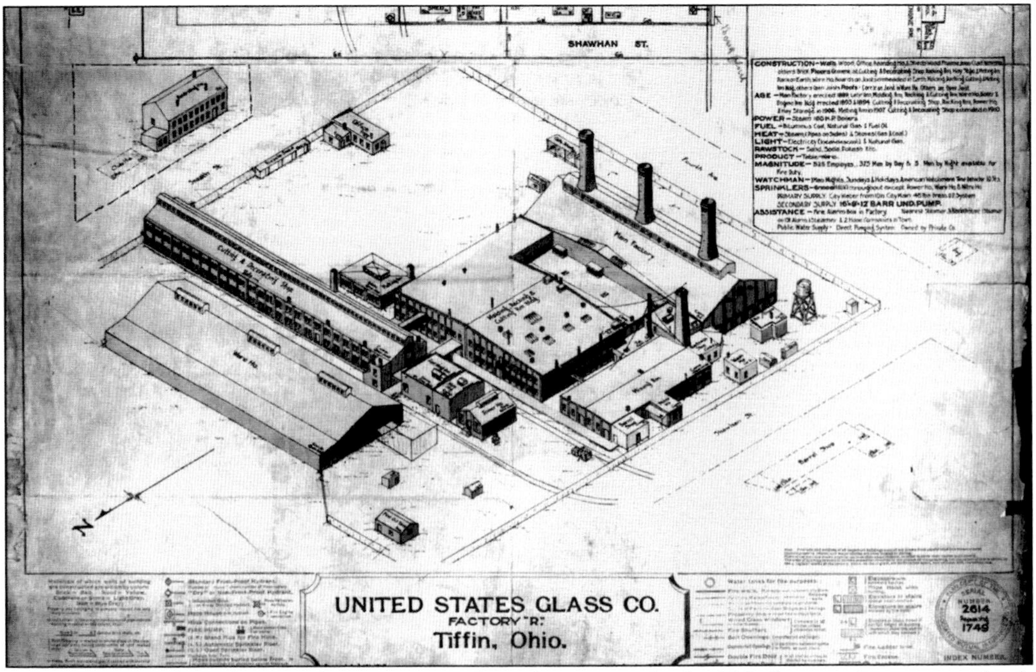

Shown is one of northwest Ohio's greatest glass plants outside the Toledo area. Its roots go back to A. J. Beatty and Sons, which moved to Tiffin from Steubenville in 1888. The original factory burned to the ground and would become part of the great glass trust (United States Glass of Pittsburgh) in 1893. Its pressed glass tableware is known as Tiffin ware throughout the world. The factory survived many setbacks until its closing in 1984.

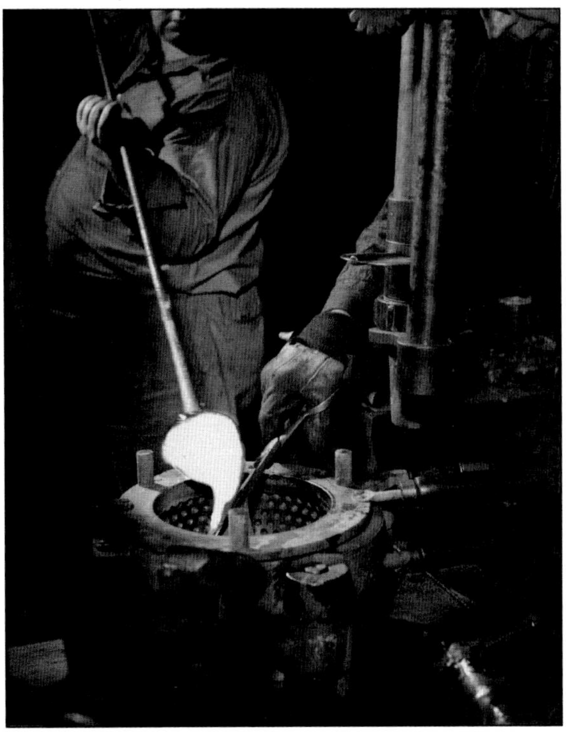

The pressing of bowls and tableware goes back to the early work of Deming Jarves of New England Glass in 1860. Another great glassmaker, William Leighton (originally of New England Glass) further improved pressed glass by introducing lime glass. While pressed lime glass lacked the weight and resonant ring of cut lead glass, it was by far cheaper for tableware.

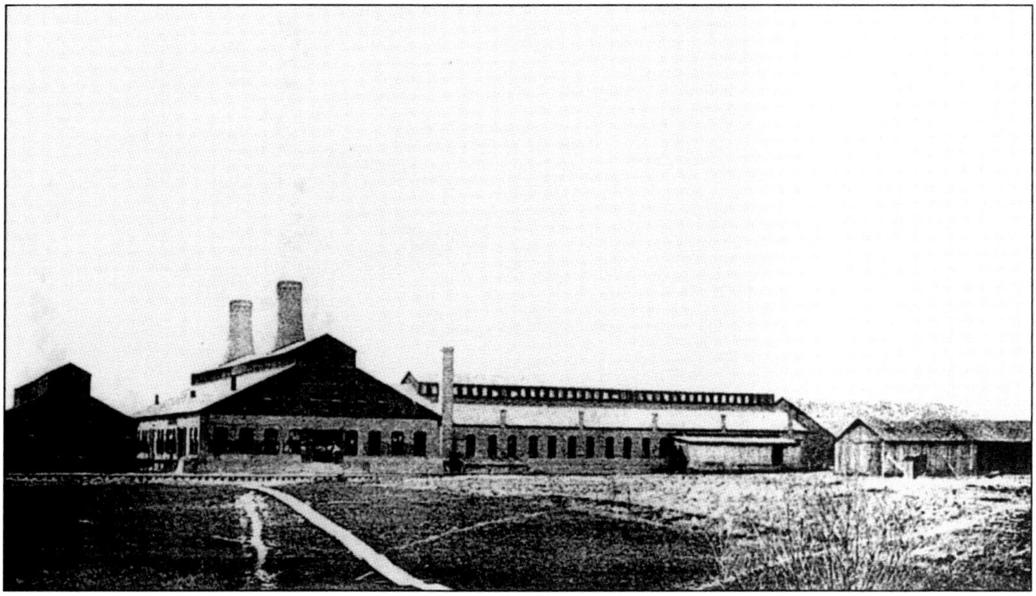

One of the first glassworks to come to northwest Ohio, this was a nationally known company in tableware and pressed stemware that moved from Bellaire, Ohio, for Findlay's free gas. Its 30-pot furnace was one of the largest in the country at the time. It ran a huge boardinghouse for over 50 boys who worked there around Davis and College Streets. In 1891, the plant became Factory M of the United States Glass Company.

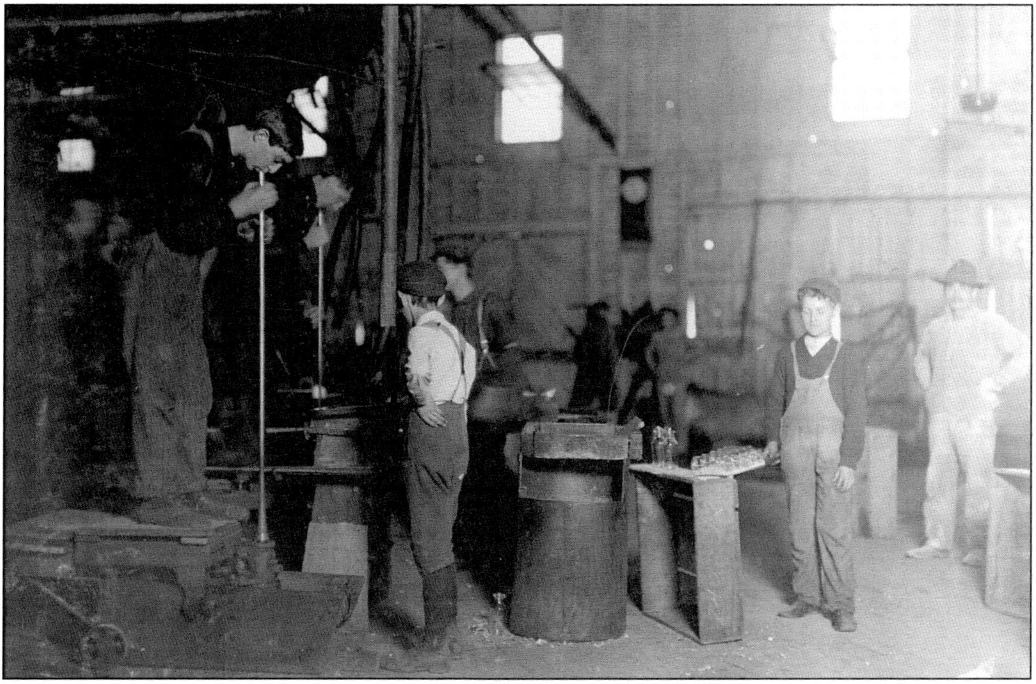

This photograph shows a small bottle-blowing operation. The carry-in boy is on the right. He carried finished bottles to be annealed. The process shown is the semiautomatic bottle mold process pioneered by Mike Owens in the 1890s. (Courtesy of the University of Toledo's Ward Canaday Center.)

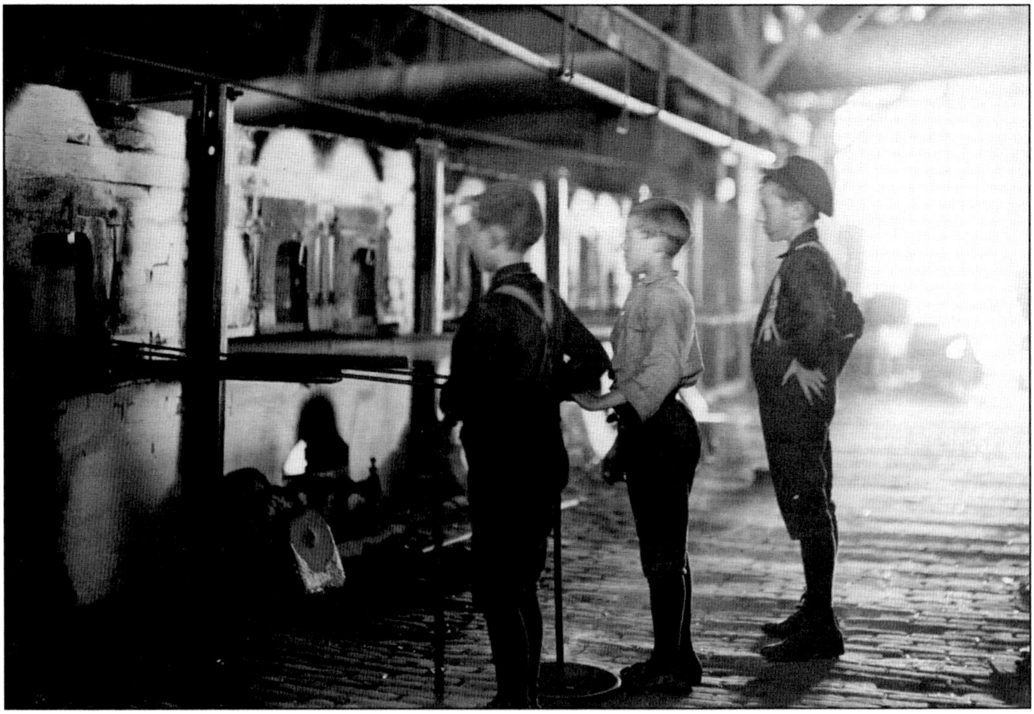

Blown glass had to be annealed (reheated and slowly cooled) to eliminate stress. The anneal or lehr was managed by unskilled boys. (Courtesy of the Hancock Historical Society.)

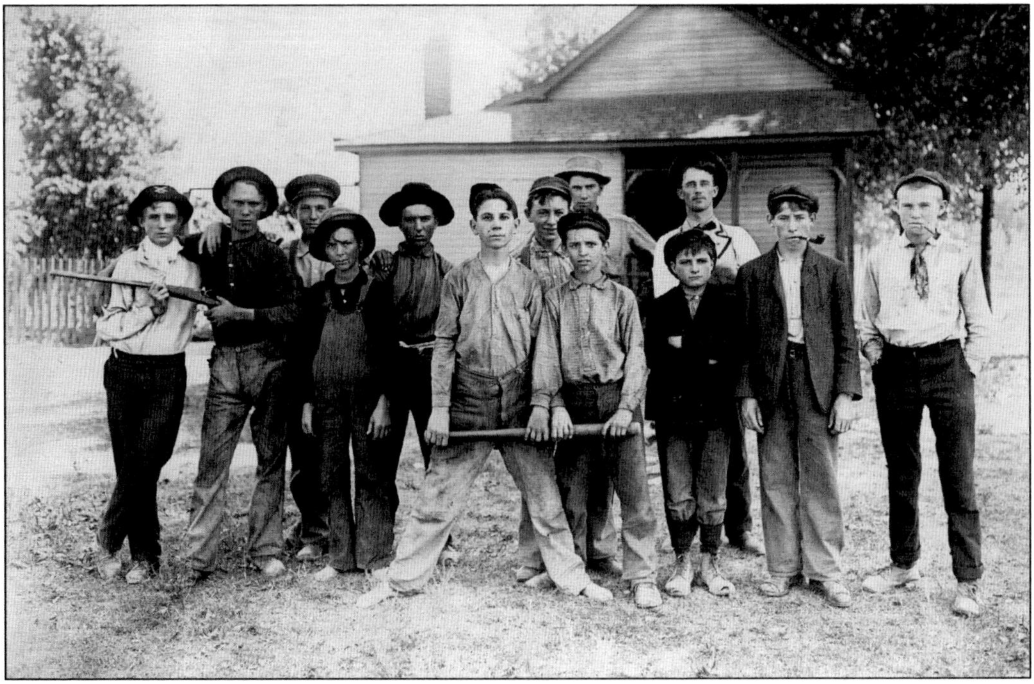

Northwest Ohio towns and companies offered boardinghouses for the glasshouse boys. A glass boy could make $3 to $6 a week, but in a boardinghouse such as the Tiffin Commercial House, they were charged $2.75 a week for room and board.

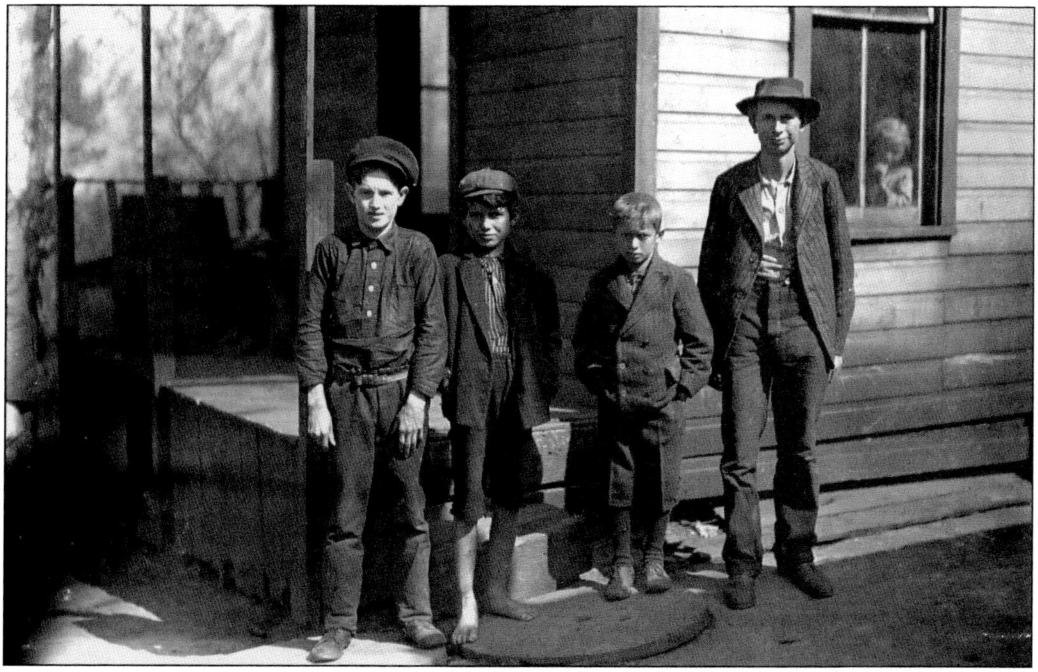

Orphans were brought in from the East to work in Ohio glasshouses. State inspectors commonly cited these boardinghouses for overcrowding, with three boys to a bed. Many advertised that they supplied the boys two clean suits, socks, and shoes. Legend has it that Mike Owens started his glassmaking career as a shoeless boy. (Courtesy of the Library of Congress.)

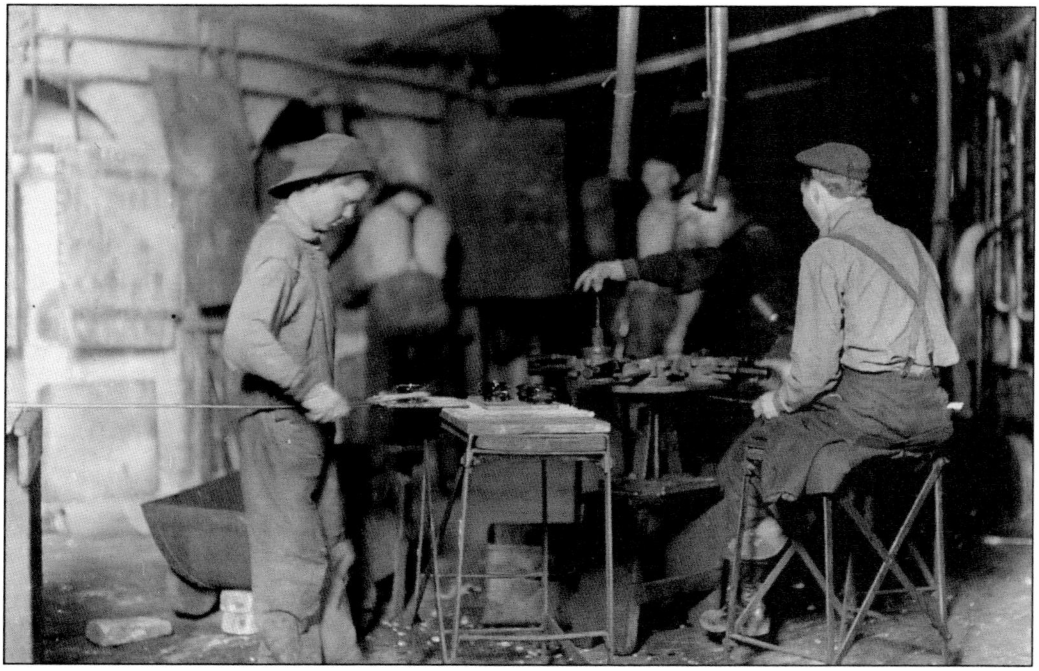

Small pressed glass works were common. These operations produced glass souvenirs, paperweights, and tableware. The special lime-soda made pressing possible. Bloomdale was a small four-pot furnace operation.

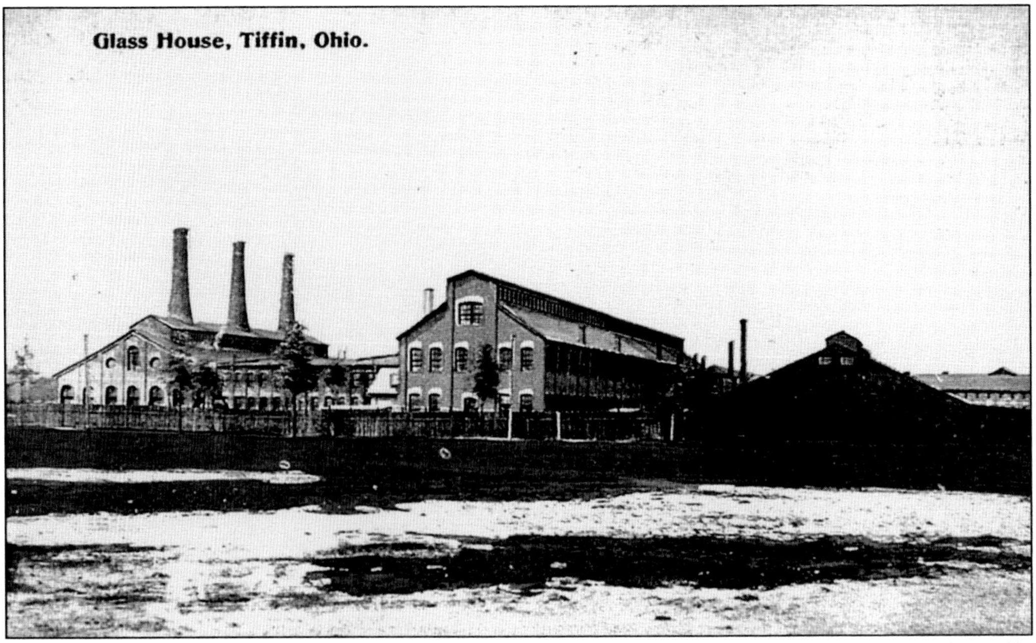

The glasshouse started in 1888 as part of A. J. Beatty and Sons, the world-renowned Steubenville glass firm. The plant had to be rebuilt many times, but it employed 600 employees and had 45 pots at its peak. The final melt was made in 1984, making it the oldest and longest-running glasshouse in northwest Ohio. (Courtesy of the Tiffin Glass Collectors.)

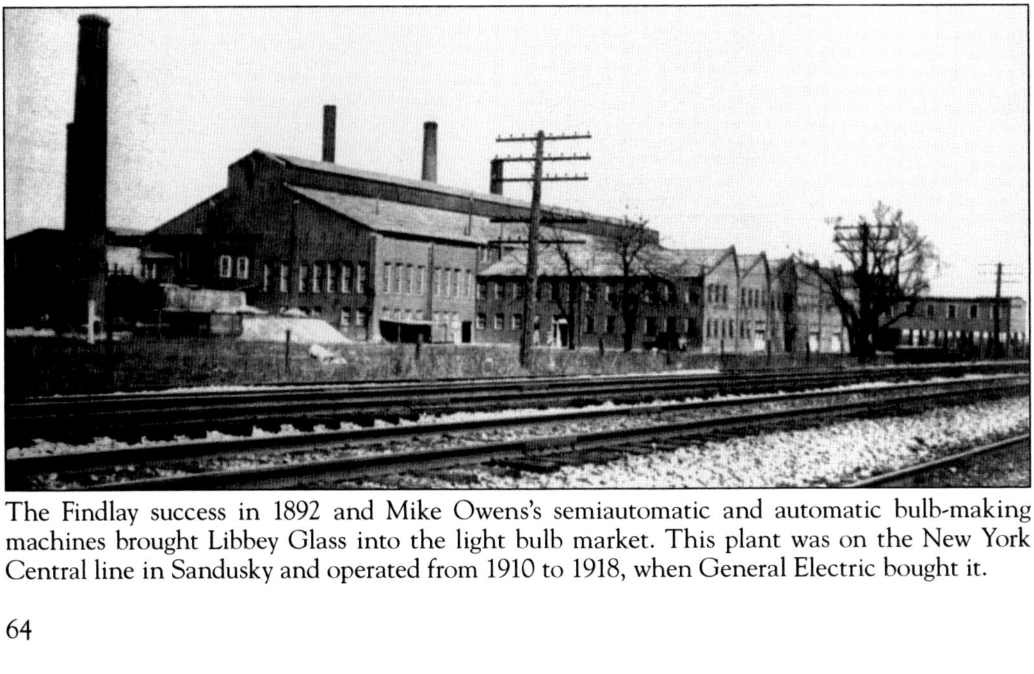

The Findlay success in 1892 and Mike Owens's semiautomatic and automatic bulb-making machines brought Libbey Glass into the light bulb market. This plant was on the New York Central line in Sandusky and operated from 1910 to 1918, when General Electric bought it.

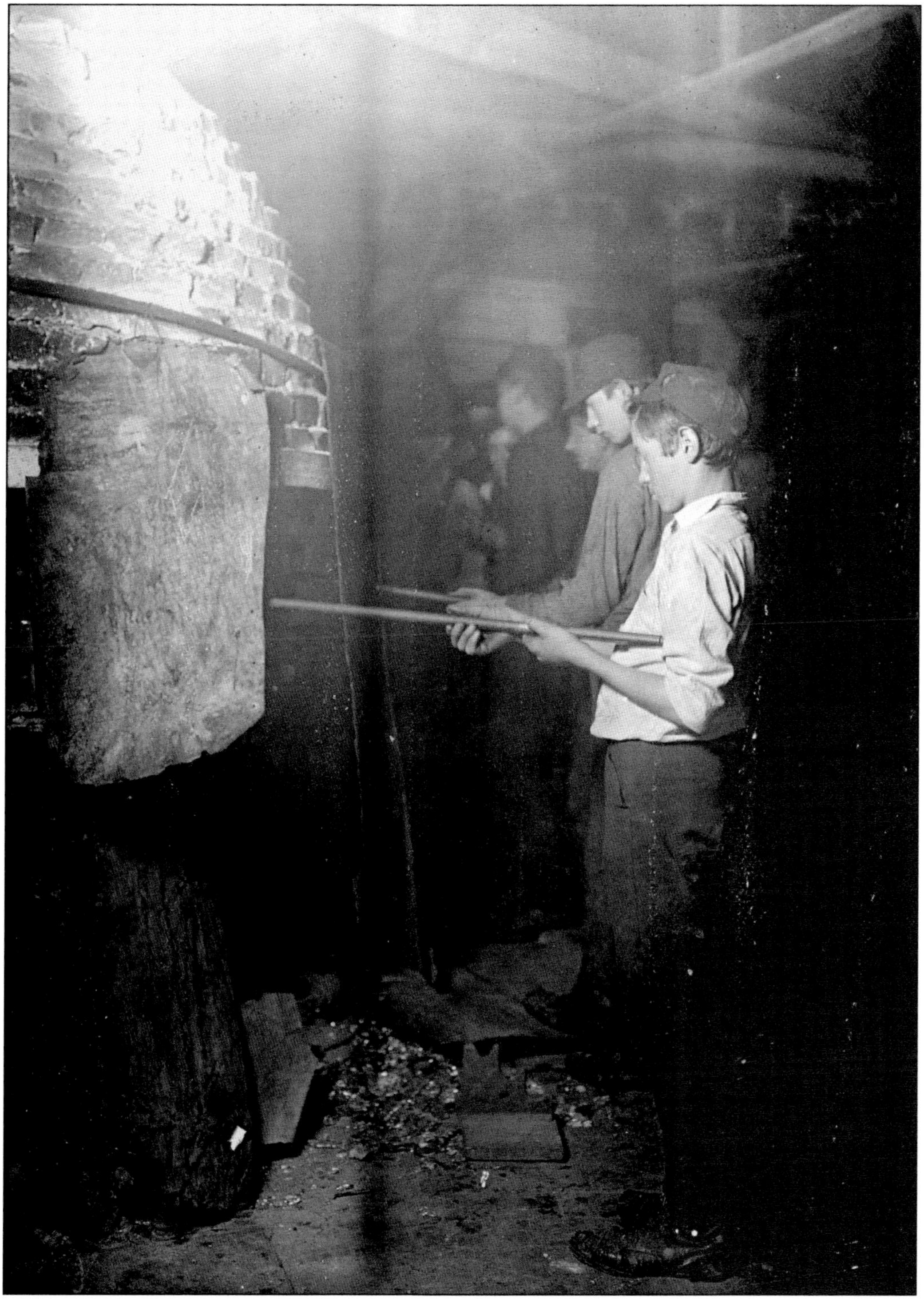

The "glory hole" was originally a small opening in the major furnace that could be used for reheating and annealing. In some plants it was a small furnace. Michael Owens started as a 10-year-old boy operating the glory hole.

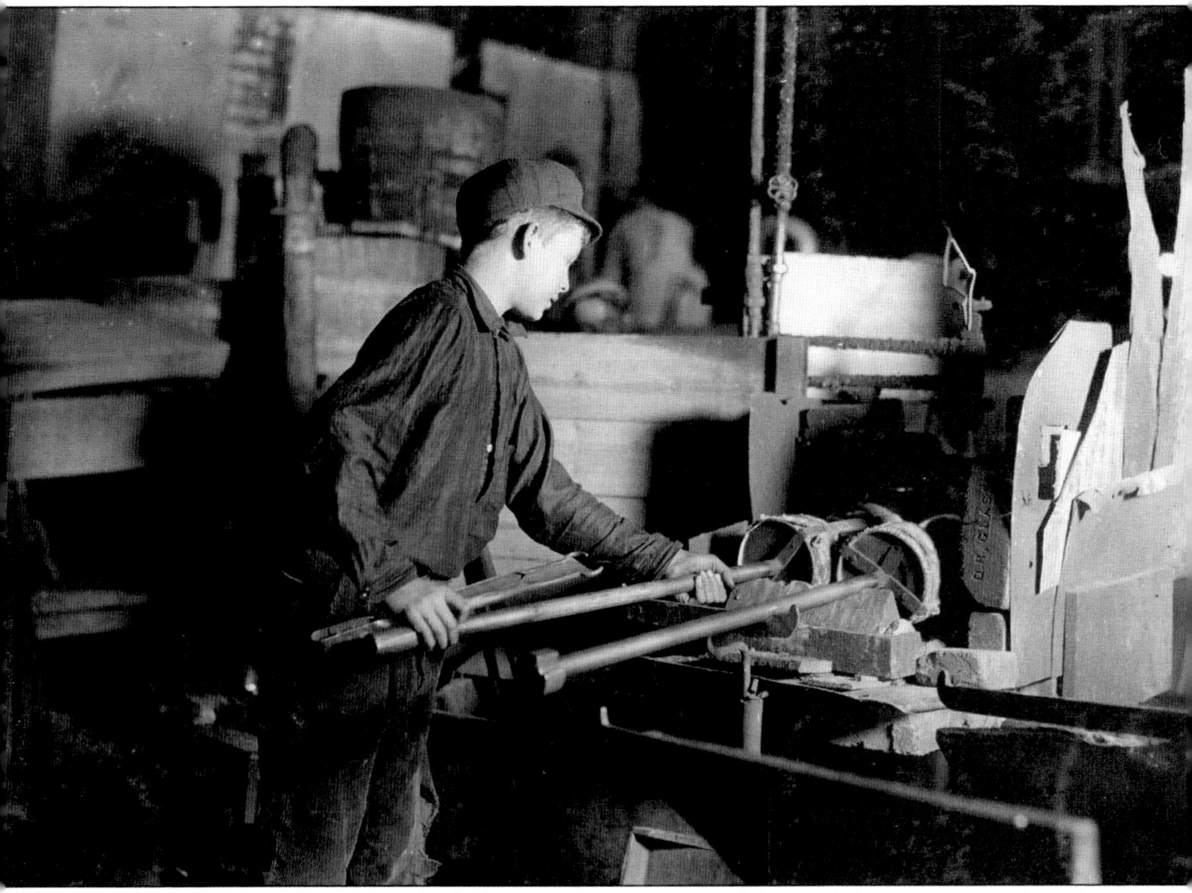

One of the most dangerous jobs for boys was the grinding of defects. There were no safety glasses or shields used.

Six

The Glass Capital

The glass industry in the 1890s consisted of the polished plate glass segment, the window glass segment, the container and bottle segment, and the pressed and blown glassware. Each segment had unique machinery, processes, and labor requirements. Different unions represented each segment as well. The window glass manufacturers actually started with a blown cylinder that was then cut and flattened. Northwest Ohio dominated the window glass market from 1890 on. Plate glass was thicker and made for larger windows; it also supplied the new automotive window market. Molten glass was cast on a flat iron table, and after cooling it, was polished and cut in pieces.

New companies came into northwest Ohio, drawing on its low freight rates and glassmaking expertise. The most important of these was Edward Ford Plate Glass. Edward Ford was the son of Pittsburgh Plate Glass's founder, Capt. John Ford. Edward had often stopped over in Toledo on his way to the southern Michigan soda ash operation. Edward concluded that the availability of quality sand and soda ash coupled with cheap transported coal made northwest Ohio the ideal spot for glassmaking. After the crop harvest in 1898, Edward started construction of Ford Plate Glass on a Wood County farm. The new plant was state of the art to produce cast-polished glass for windows. The heating fuel would be producer gas, a by-product of coal processing. In addition, the plant was one of the first to use electricity extensively. The plant was extremely successful and ultimately would become the anchor of Libbey-Owens-Ford in 1930 (today it is Pilkington).

Many of the window glass companies had remained in the area after the gas boom, and these created new industries such as cut glass companies. A number of cut glass companies moved into Findlay, Bowling Green, and Toledo from 1900 to 1920. Fostoria and Sandusky had light bulb assembly plants supplying General Electric and the replacement market. Libbey Glass again dominated the 1904 world's fair at St. Louis with its cut art glass. With Libbey and Ford Glass leading, northwest Ohio was dominant in the glass industry.

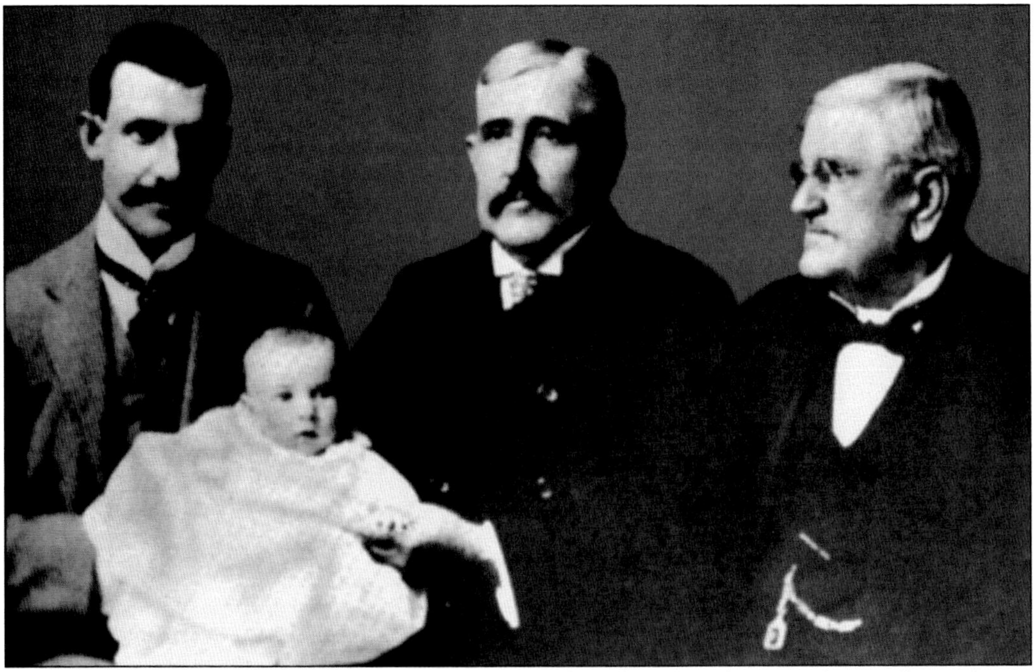

From left to right are John B. Ford, John B. Ford III, Edward Ford, and Capt. John Ford. Northwest Ohio could not claim the title of Glass Capital until the arrival of Ford Plate Glass in present Rossford in 1899. The Fords split from Pittsburgh Plate Glass, which the captain had cofounded. (Courtesy of the Hancock Historical Society.)

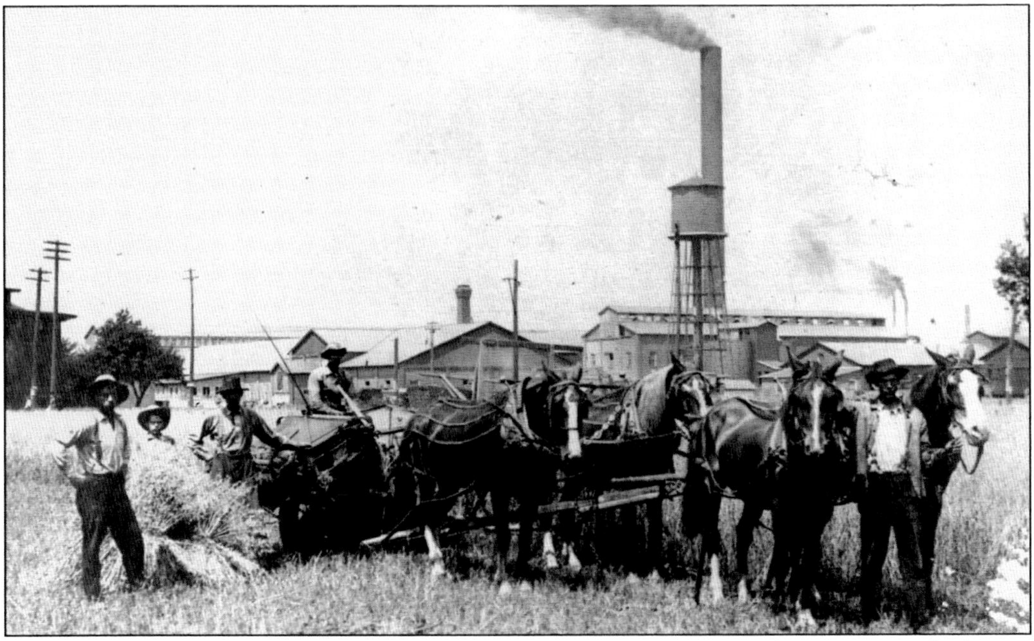

Ford Plate Glass was located on a working wheat farm for many years. The name Rossford came from Edward Ford's wife, Caroline Ross. The plant cast, ground, and polished thick plate for large windows. The Ford Plate Glass plant would be the largest in the world in 1900, producing three million square feet of plate a year. (Courtesy of the Rossford Public Library.)

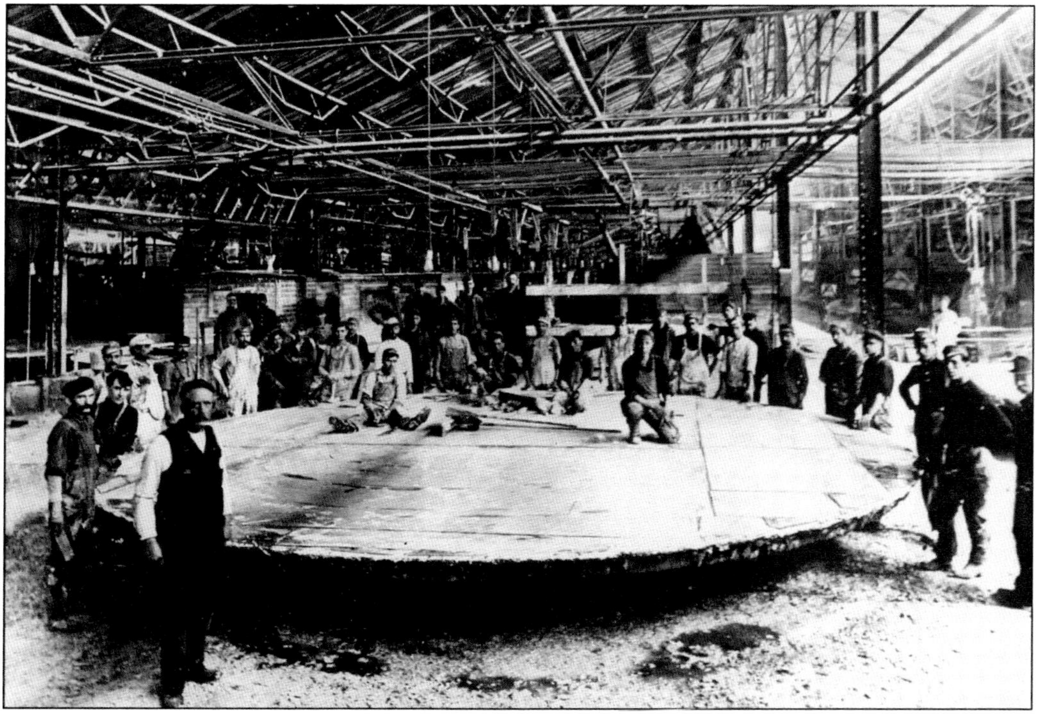

The grinding and polishing of glass was a labor-intensive process in the first 20 years of Ford Glass. Ford would be one of the first to install an automated grinding and polishing line in the late 1920s.

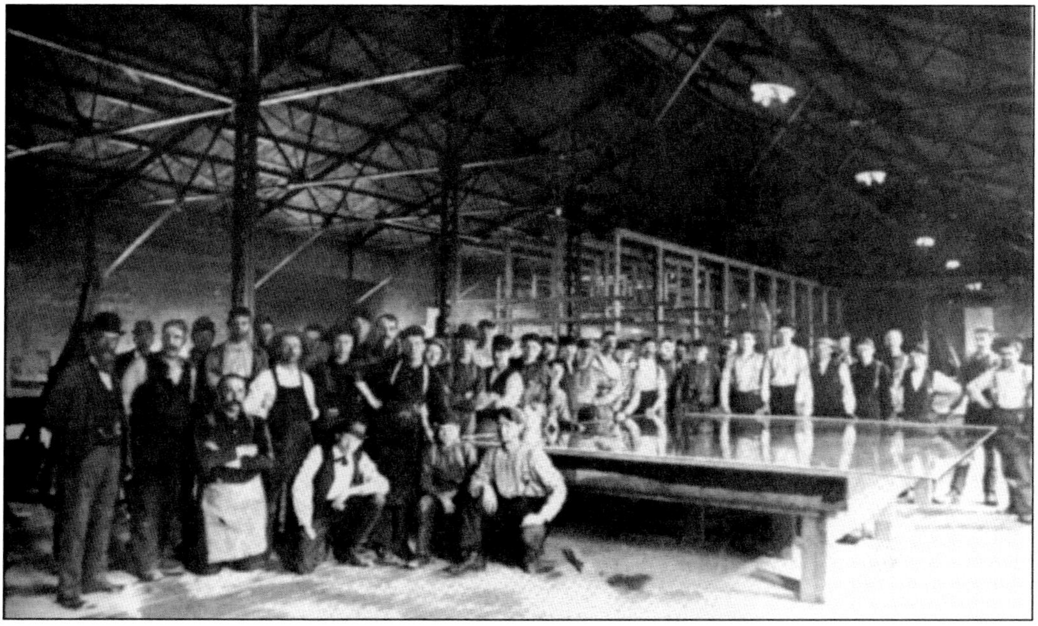

At the extreme left is Matthias Pepper, who came with Ford from Pittsburgh Plate Glass in 1899. Pepper was the first to develop the methods for the production of polished plate glass at England's Pilkington Brothers. He was the first plant manager at Rossford. (Courtesy of the Rossford Public Library.)

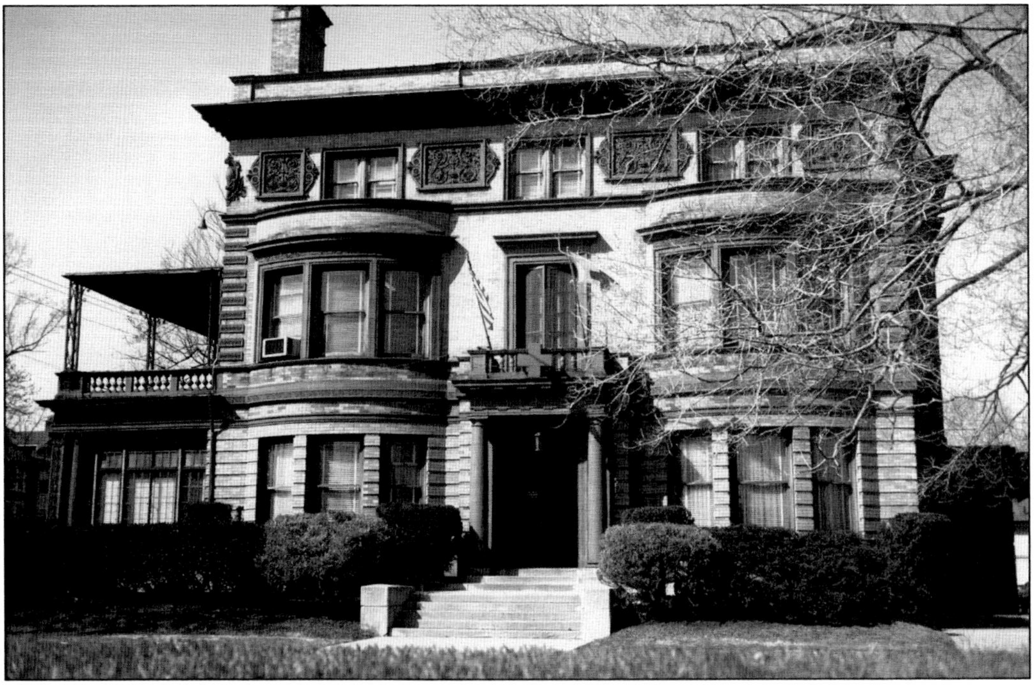

Edward Ford built a town around his new factory. The town was named Rossford after his wife, Caroline Ross. (Courtesy of the Toledo Lucas County Library.)

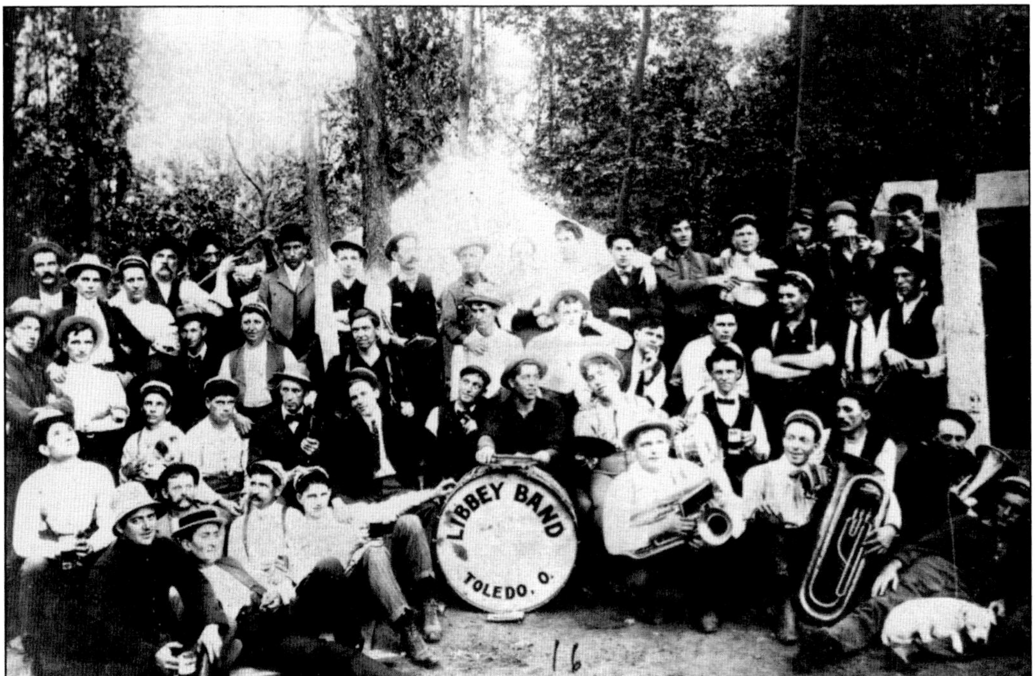

In the early 1900s, Clark Lake was a popular spot for Libbey company picnics. Michael Owens also had a cottage at Clark Lake in the Michigan Irish Hills area.

The cover shows the cutting of the famous cut Libbey punch bowl for the 1904 St. Louis world fair. The 134-pound bowl was the largest cut-glass piece in the world. It was considered a major technical feat. The cutter was one of the world masters, John Rufus Denman.

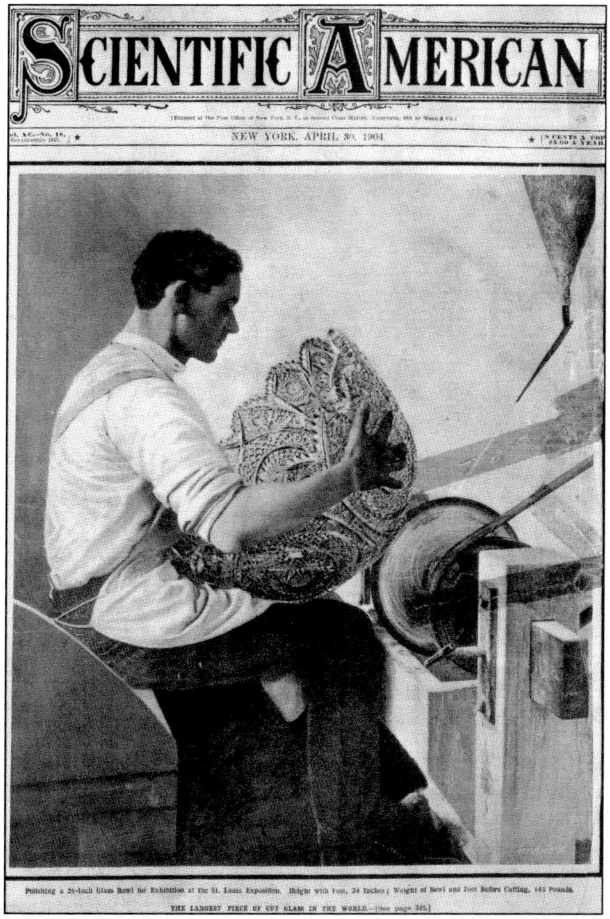
SCIENTIFIC AMERICAN

NEW YORK, APRIL 30, 1904

Polishing a 25-Inch Glass Bowl for Exhibition at the St. Louis Exposition. Height with Foot, 24 Inches; Weight of Bowl and Foot Before Cutting, 143 Pounds.

THE LARGEST PIECE OF CUT GLASS IN THE WORLD.—[See page 348.]

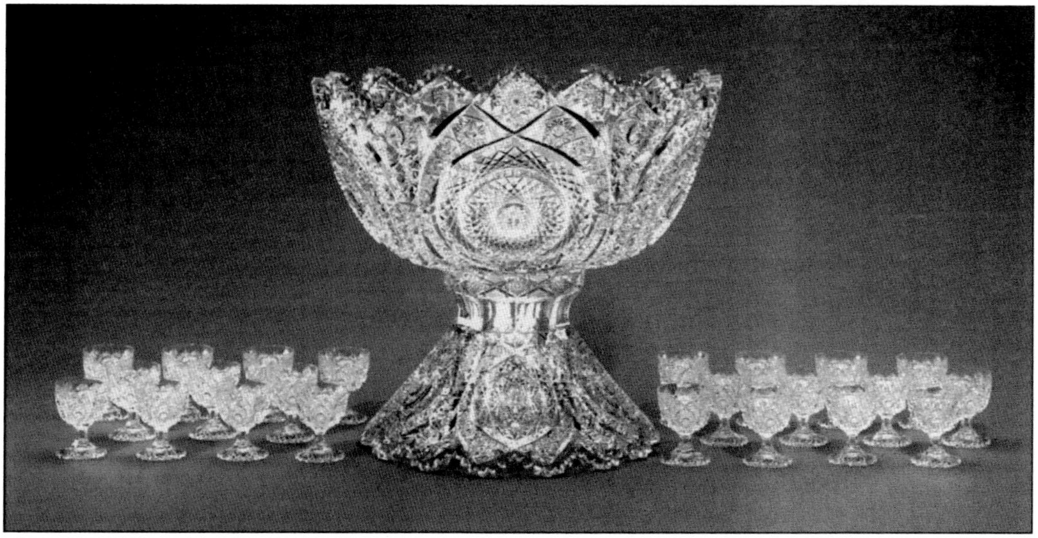

This punch bowl by Rufus Denman won a citation of honor at the 1904 St. Louis fair. It was similar to one made a few years earlier for Pres. William McKinley. (Courtesy of the Hancock Historical Society.)

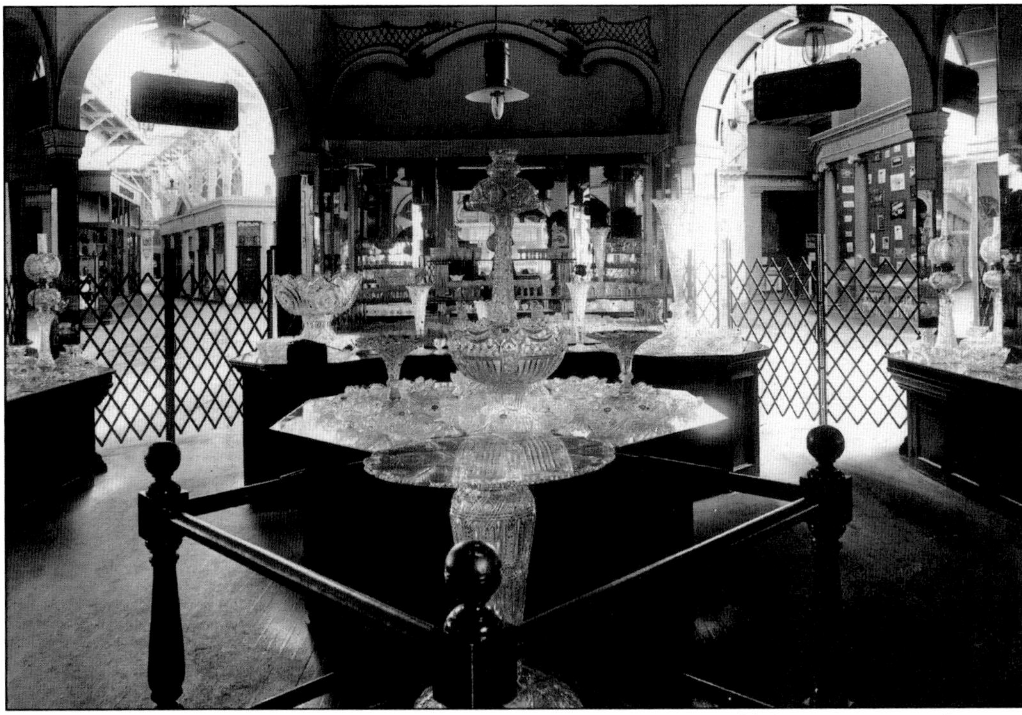

The punch bowl on the left was made to celebrate Cresceus, a world-famous trotting horse of George Ketcham. Master craftsman John Rufus Denman made the glass table in the forefront. The floor and table lamps shown were also high-end products of Libbey Glass. Libbey's glassblower C. Nolan was world famous for these blown lamps.

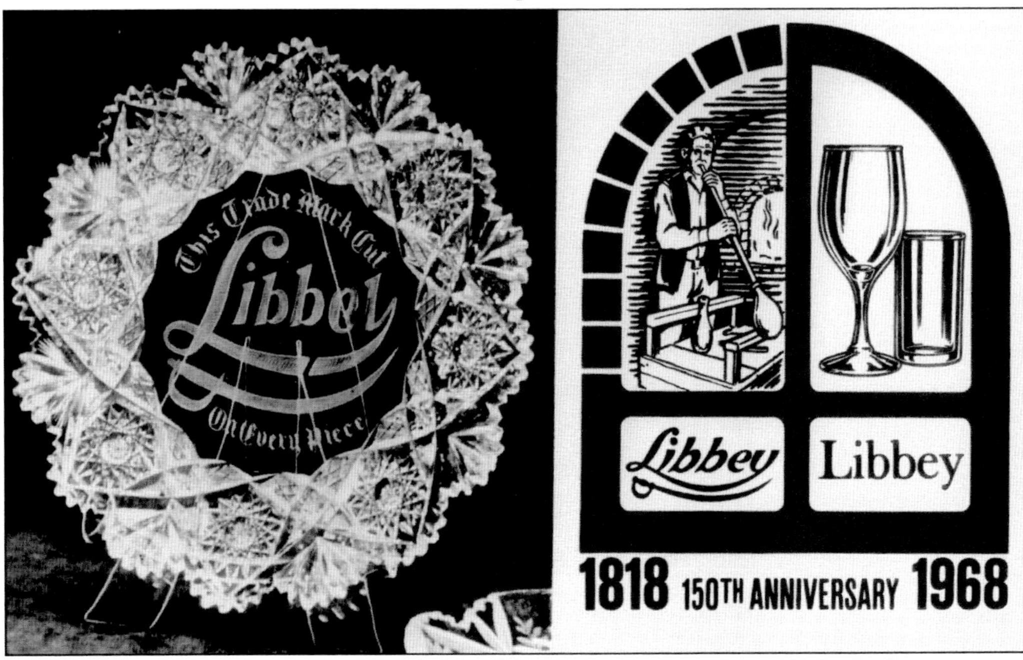

The plate at the left was made to celebrate Libbey's new trademark in 1896, which was put on pieces until 1906. (Courtesy of the University of Toledo's Ward Canaday Center.)

Starting in a small wooden building, Edward Libbey formed Owens Bottle to be an experimental operation for Owens. (Courtesy of the University of Toledo's Ward Canaday Center, the Paquette Papers.)

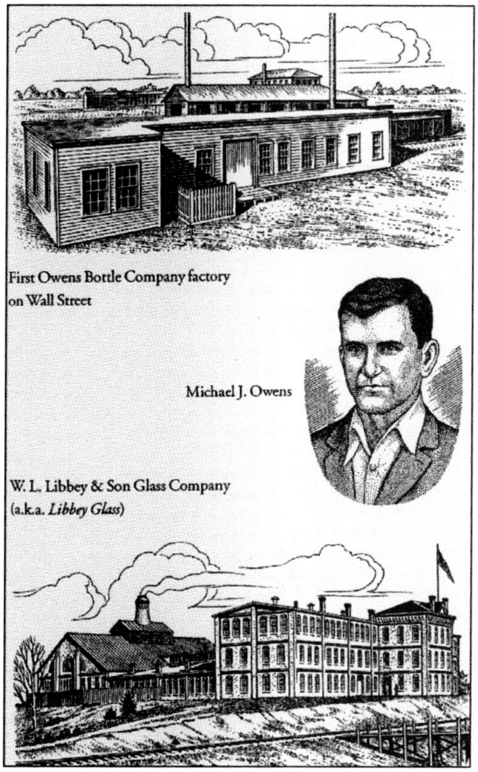

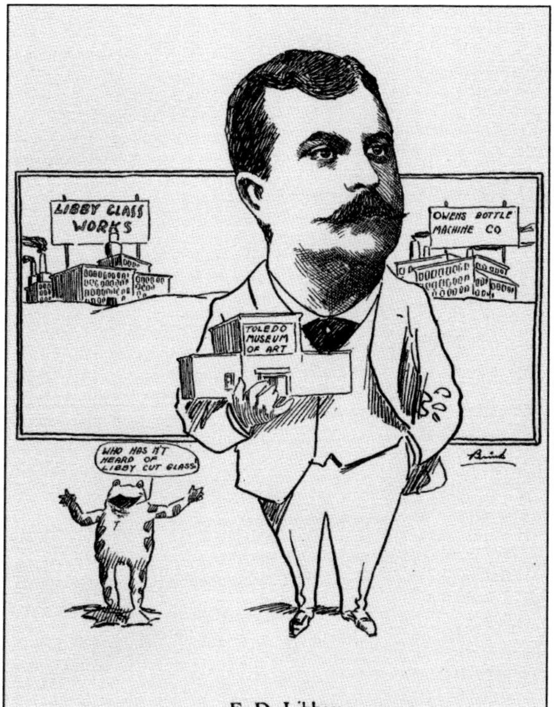

E. D. Libbey
Of the Libbey Glass Company and The Owens Bottling Machine Company.

Edward Libbey was invested in art glass, cut glass, bottle equipment and machinery, light bulb manufacture, and glass tubing through various companies. Libbey formed separate companies to avoid the trust busters in Washington.

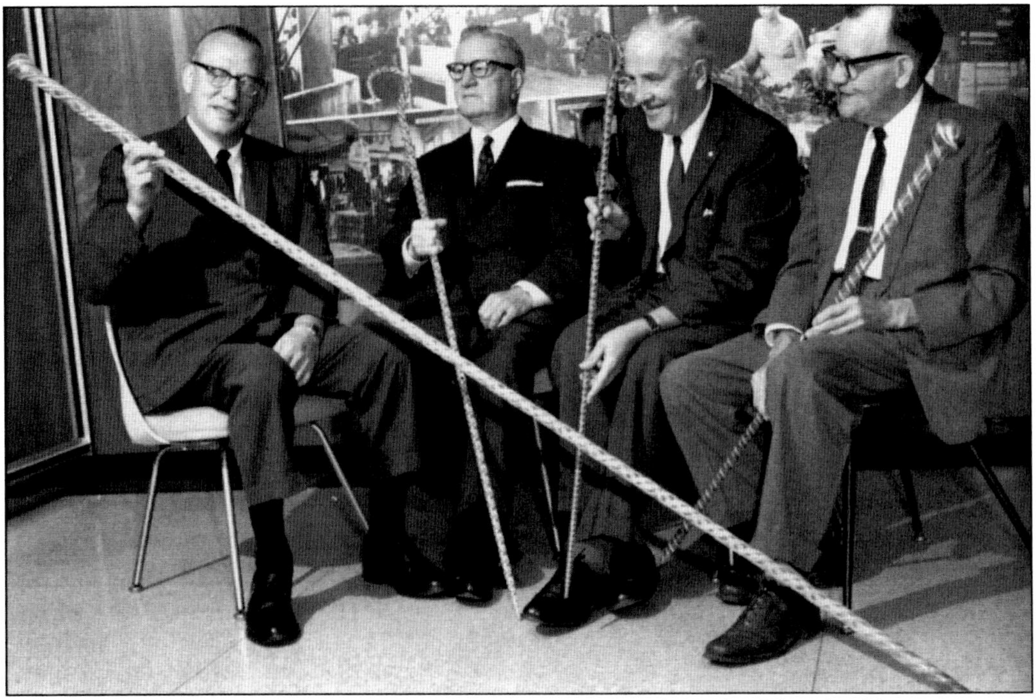

The glass cane made from excess glass in the pot was the trademark of the glassblower.

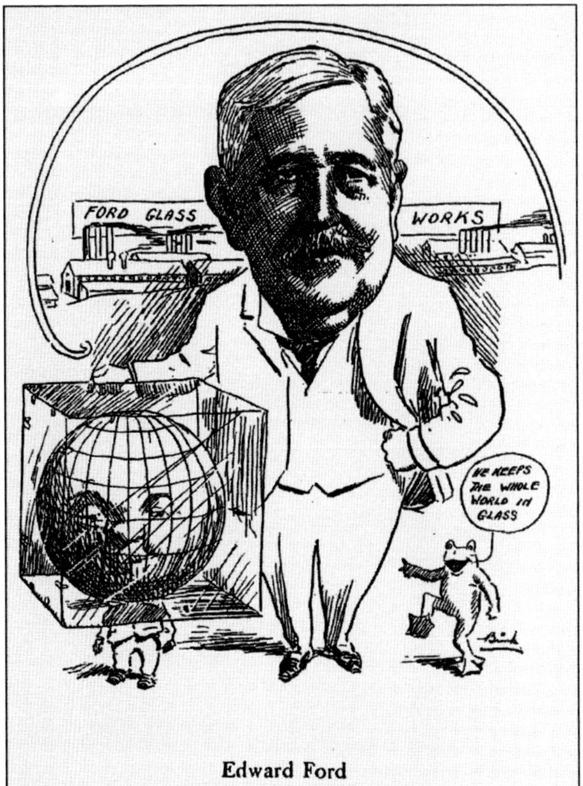

Edward Ford
Of The Ford Glass Company.

Edward Ford stands with Libbey and Dalzell as the troika of industrialists that made northwest Ohio the glass capital. He became an investor in Owens Bottle early on and was on the board of Toledo Glass in 1914. Ultimately Ford Glass would merge with Owens Bottle to become Libbey-Owens-Ford.

Seven

A Revolution in Bottle Making

For over 3,000 years, bottles and cylindrical blown glass had been made the same way it was being made in 1891. Bottles were hand blown using the Venetian method; the mold improvement was advanced by Owens. The mold was prepared, opened, closed, and cleaned by a mold boy. A shop with six men and boys could produce 2,000 to 3,000 bottles, chimneys, and bulbs in 24 hours at a cost of $1.80 a gross. A mold boy strike at the Libbey Glass Findlay bulb plant motivated Michael Owens to develop a semiautomatic mold-blowing process. The system allowed the glassblower to use a foot pedal to open and close the mold. Edward Libbey formed Toledo Glass in 1895 to allow Owens to pursue automation in glassmaking.

Owens had developed a handheld suction device from a bicycle pump to form crude bottles. With the formation of Toledo Glass, he was given a trained engineer, Emil Bock, as an assistant. Their first true success was a four-ounce petroleum jelly jar in 1899. The long neck of beverage bottles required extensive process development to overcome this bottleneck. The first Owens machine could produce 17,000 to 18,000 bottles using two machine operators in a 24-hour period at a cost of 10¢ a gross.

Owens was interested in manufacturing bottles, but Libbey and his two lawyer investors, Clarence Brown and Fredrick Geddes, were the real money in a horizontal monopoly of leasing machines. In 1903, Owens Bottle Company was formed by Libbey, Owens, Walbridge, Brown, and Geddes to lease the machines. One advantage of automated bottle production over handblown was uniformity. The uniformity of bottles, especially prescription bottles, allowed the government to pass in 1914 the Pure Food and Drug Administration Act, which guaranteed a specific measure of product.

Owens pioneered the use of standardized parts and automated conveyor systems far before Henry Ford. He developed "platform" manufacturing decades ahead of the automotive industry. He developed a quick mold change machine to adapt to the customization of bottles demanded. He worked with his supply chain, such as the railroads, to develop better and more efficient materials handling for glass plants.

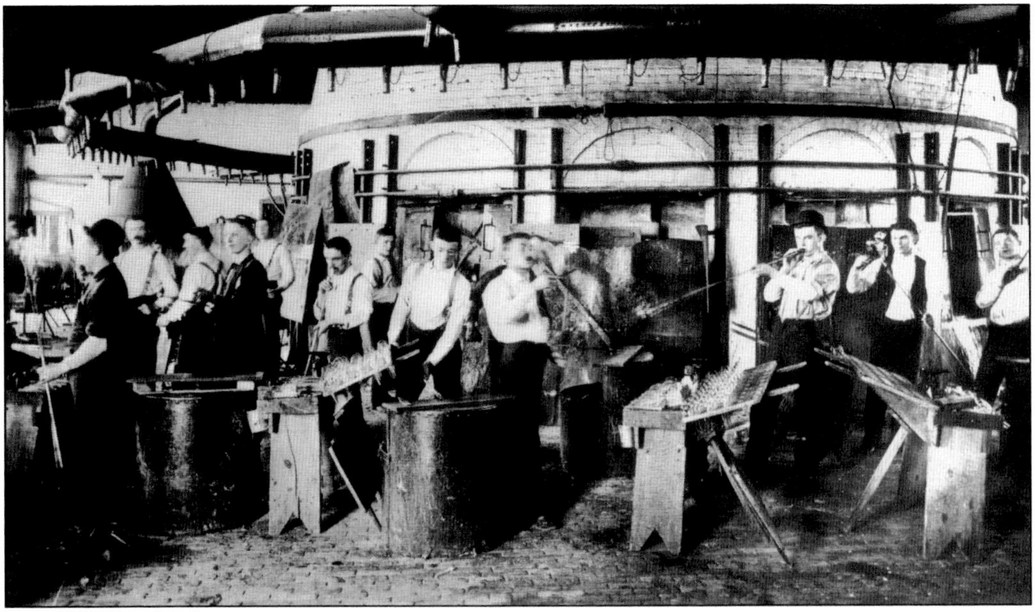

Hand blowing bulbs required higher production rates, which were much higher than tableware or art pieces. Libbey Glass, while one of the first producers of Edison glass bulbs, had a very small part of the market when Libbey came to Toledo.

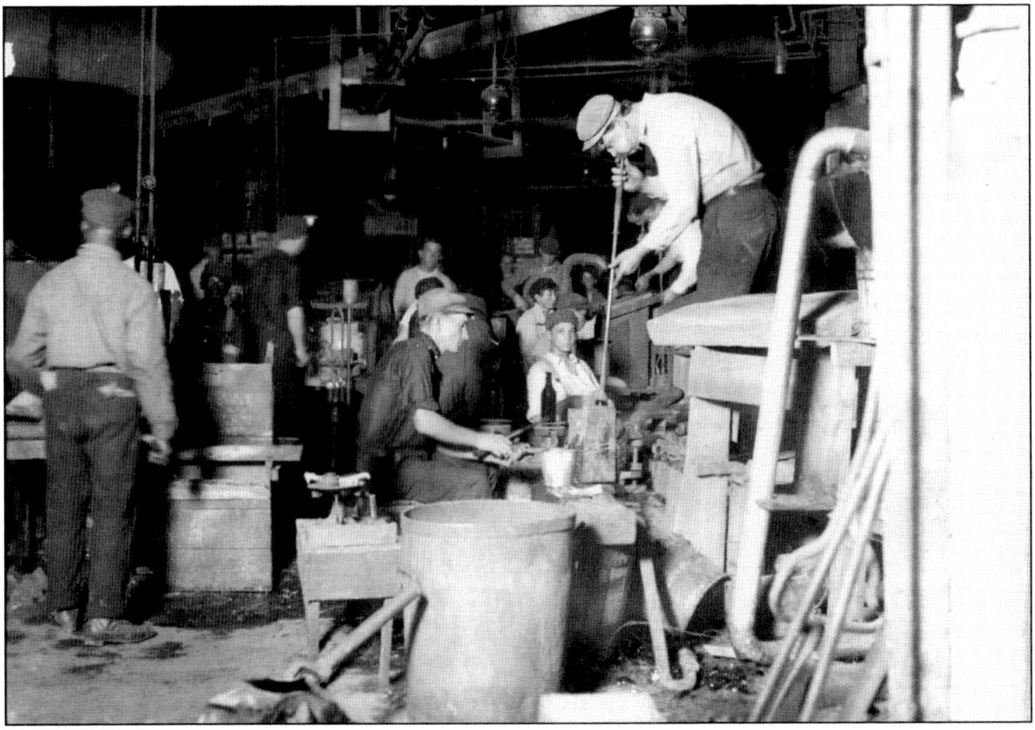

As early as 1870, glasshouses were using wooden molds to help shape bottles. The mold boy worked with the glassblower to produce a bottle. (Courtesy of the Hancock Historical Society.)

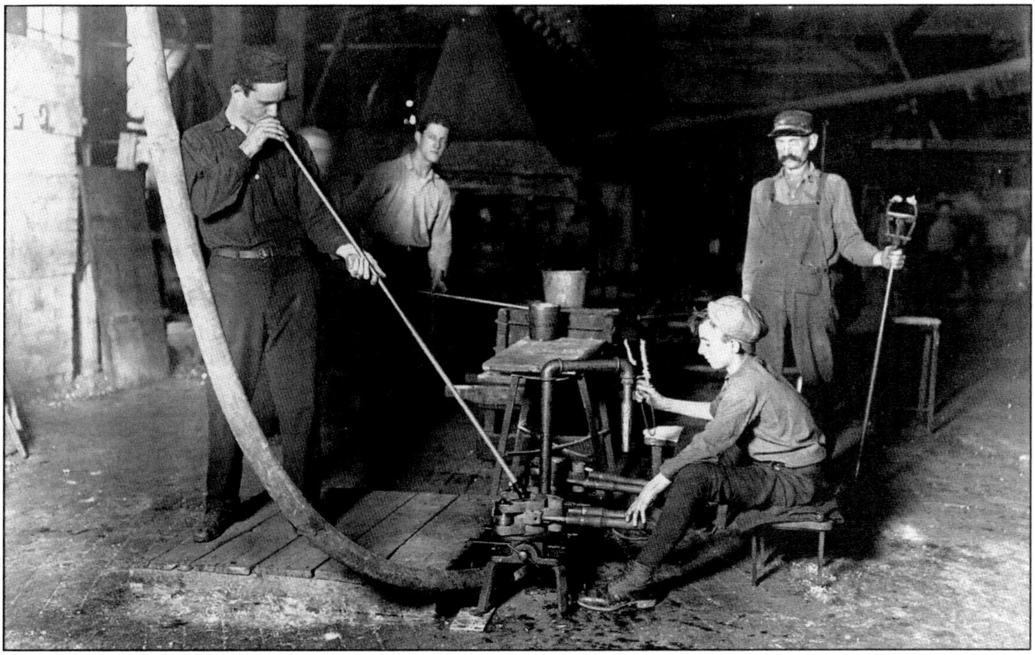

The first improvement in handblown bottles evolved in the 1870s. A bottle mold was held by a boy as the glassblower formed the bottle in the mold. Michael Owens had been a mold boy at Hobbs-Brockunier Company in Wheeling, West Virginia. Like so many innovative glass companies, a former New England Glass Company employee founded Hobbs-Brockunier. A hand blowing crew consisted of three skilled men and four boys.

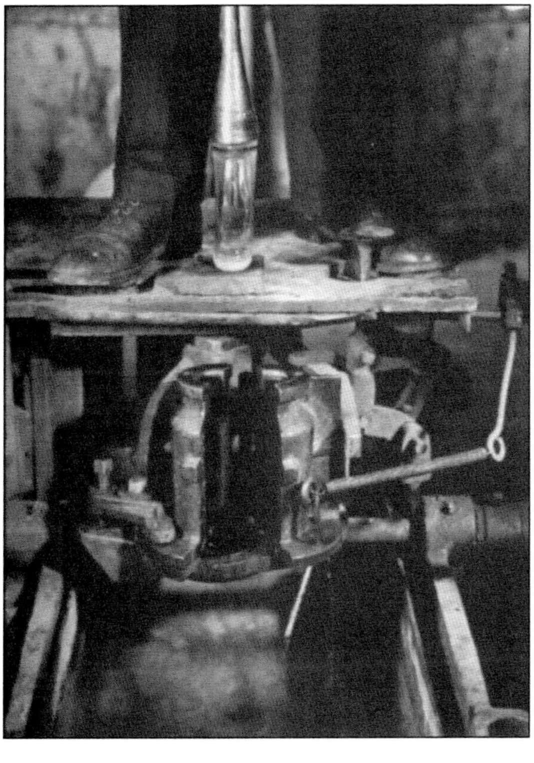

New England Glass had pioneered the use of the bottle mold in the 1840s. The molds were cast iron and coated with a paste to eliminate sticking. Pressing molds for tableware were made of copper or brass and could allow for detailed surface patterns. (Courtesy of the University of Toledo's Ward Canaday Center.)

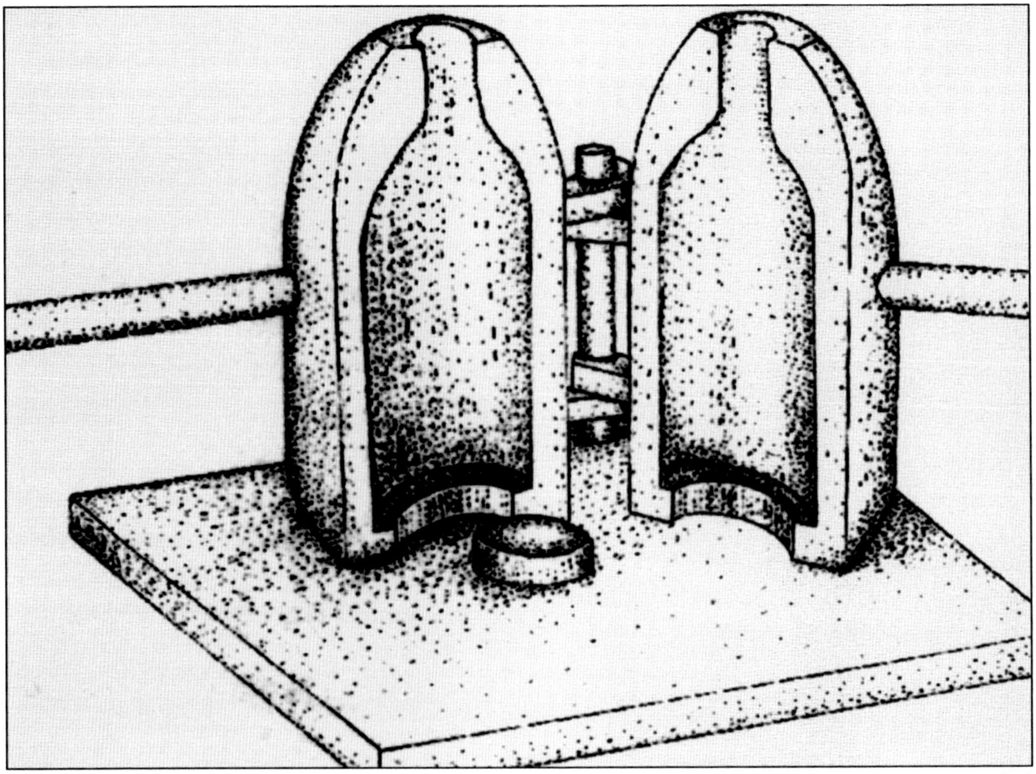

This variation allowed for a nameplate bottom to be exchanged. Such variations were needed because of the expense of machining these molds of cast iron.

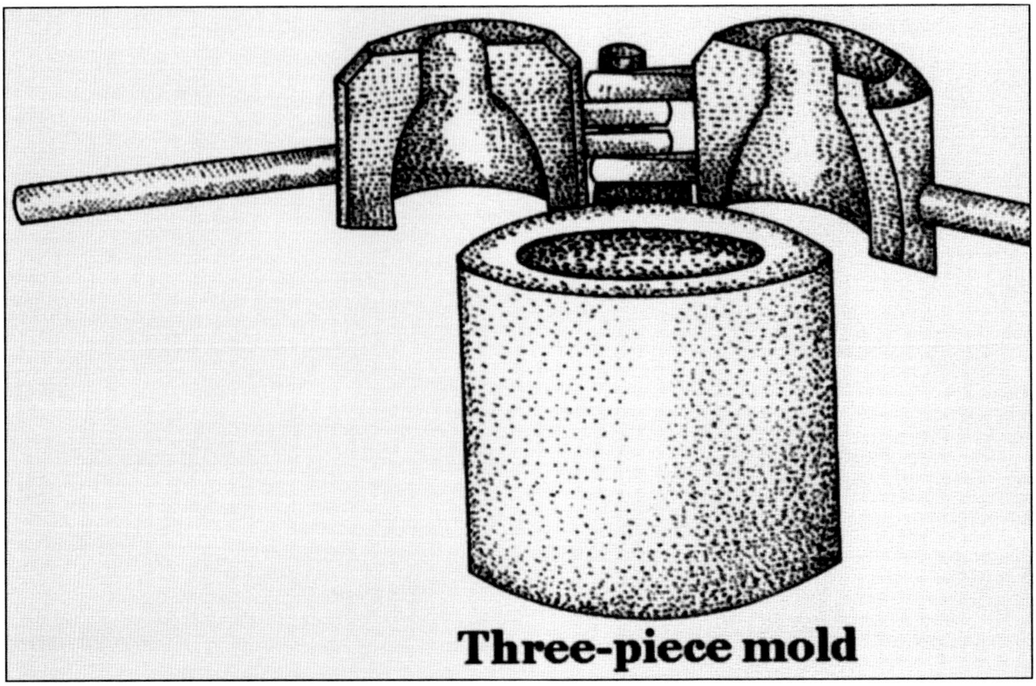

The three-piece mold allowed for more variation in bottle size and shape.

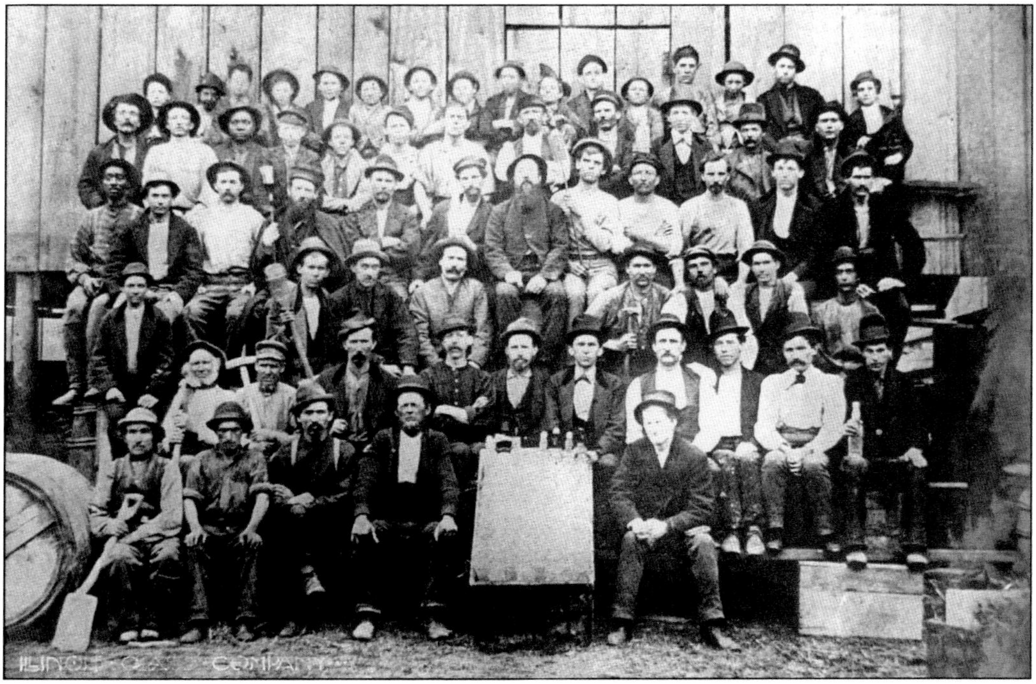

Illinois Glass was the nation's largest producer of bottles and containers in the last half of the 19th century. In the early 1900s, Illinois Glass was a major user of the Owens bottle and would ultimately merge with Owens Bottle to form Owens-Illinois in 1929.

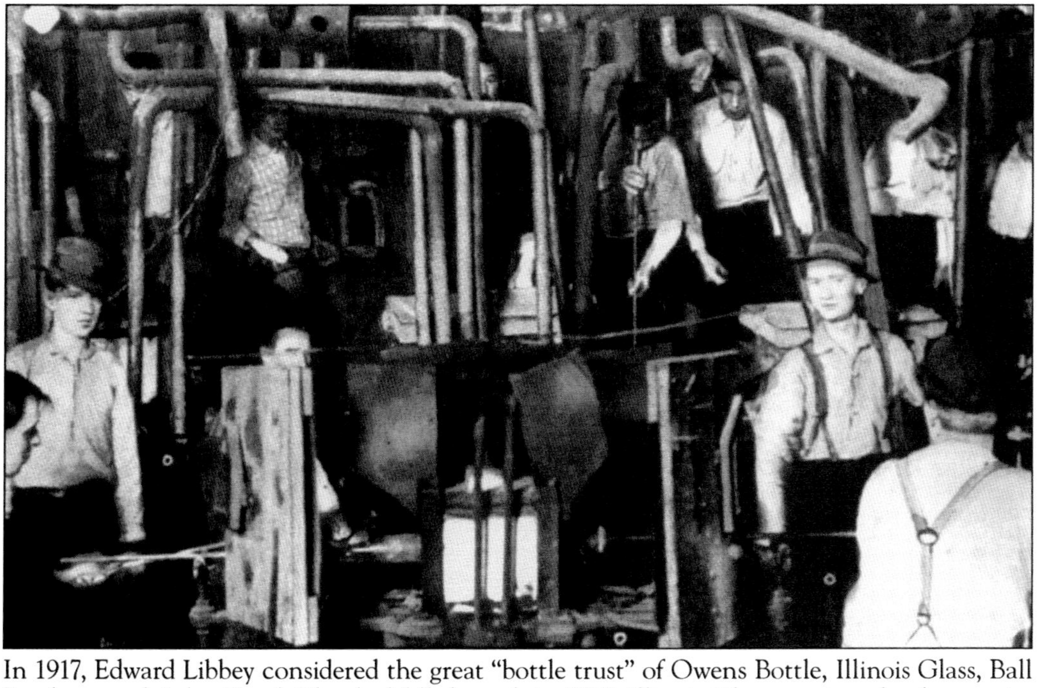

In 1917, Edward Libbey considered the great "bottle trust" of Owens Bottle, Illinois Glass, Ball Brothers, and Atlas-Hazel. The deal fell through in 1920. Illinois Glass continued to be a major factor in bottles until it successfully merged with Owens Bottle. (Courtesy of the University of Toledo's Ward Canaday Center.)

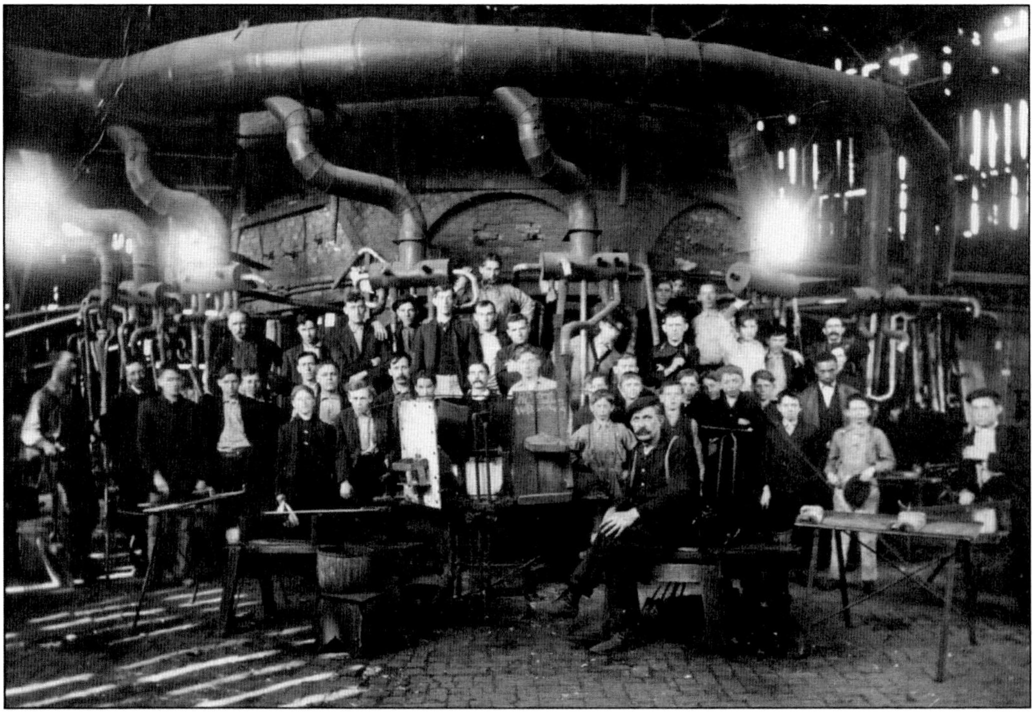

By 1902, Illinois Glass was America's major producer of liquor bottles and a major glass producer. It would have exclusive rights to the Owens bottle machine for liquor bottle production.

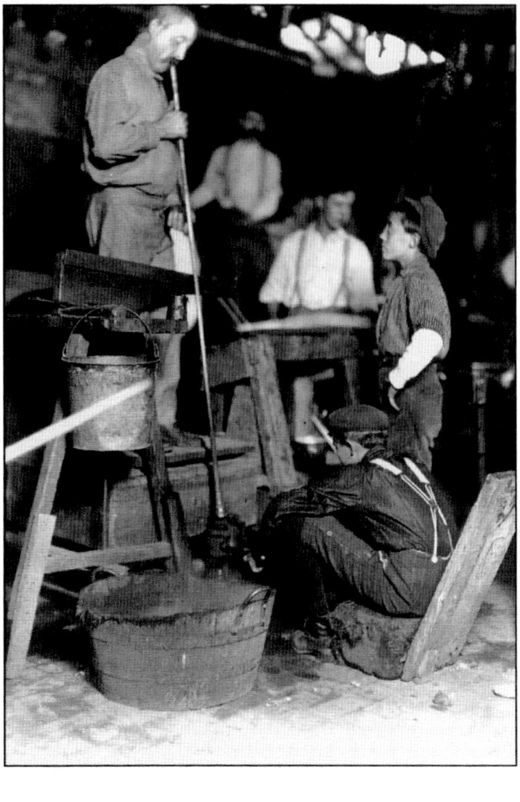

The mold boy not only opened and closed the glass bulb mold but also washed the mold with a carbon resin. The mold boy rested on the home chair as seen. The other boy was known as a cut-off boy or carry-in boy, who took the tumbler to an annealing furnace known as a lehr.

The next step in bottle and bulb making was the foot-operated mold. Facing a strike by mold boys at the Libbey Glass bulb plant in Findlay motivated Mike Owens to invent an automatic mold, eliminating the need for mold boys. (Courtesy of the University of Toledo's Ward Canaday Center.)

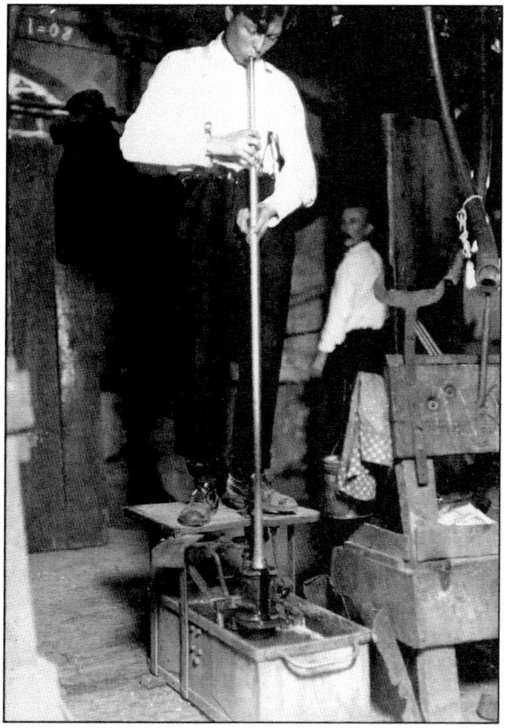

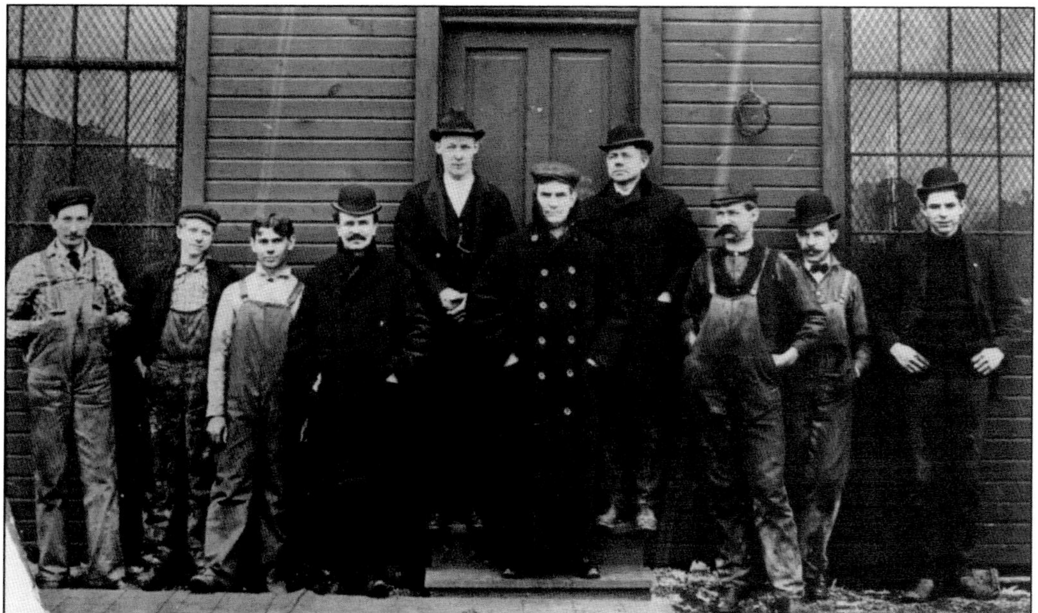

Owens and Edmund Libbey formed Toledo Glass in 1895 to develop an automatic bottle-making machine. Libbey gave the small company a wooden machine shop behind the Libbey Glass plant. Owens is in the center of the photograph with superintendent Emil Bock and assistant Bill Schwenzfeier at his sides. Bock would leave before the final bottle machine was perfected, but he would go on to invent many things, including the tapered ball bearing. Richard LaFrance, future chief engineer, is at the extreme right.

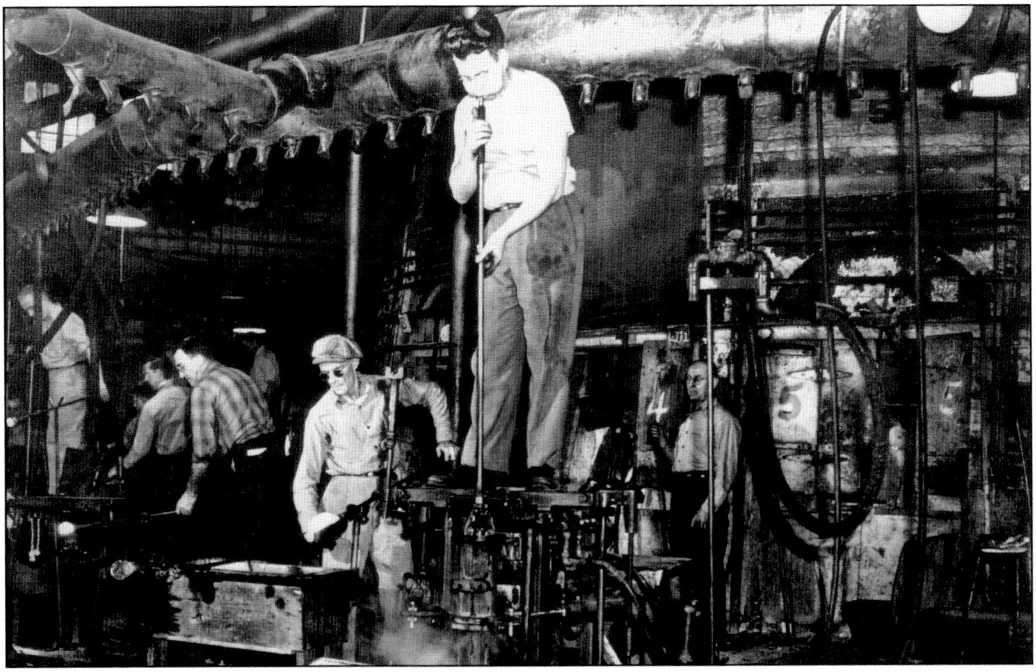

Between 1895 and 1900, Toledo Glass and Michael Owens perfected the semiautomatic process. Here is a glassblower and gatherer—the mold boy has been eliminated with the semiautomatic process. (Courtesy of the University of Toledo's Ward Canaday Center.)

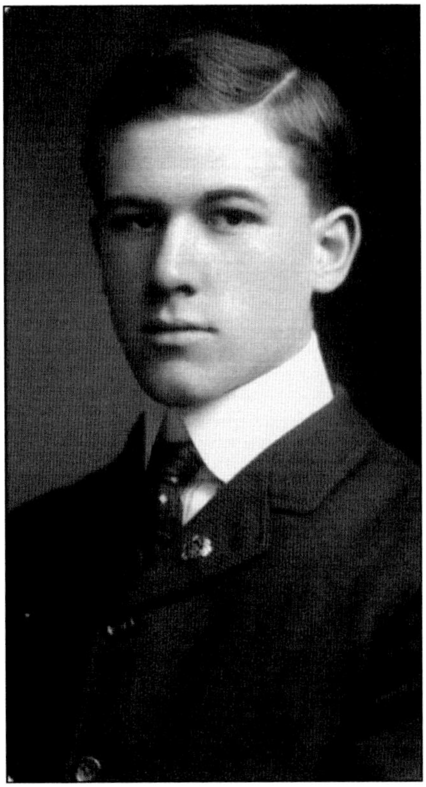

Richard LaFrance joined the Owens development in 1901. He was only 17 but had completed his study of mechanical drawing at old Central High in Toledo. He was part of a European start-up for Owens Bottle's first plant in Manchester, England, in 1907. In 1910, after Emil Bock resigned, LaFrance took over as Owens's chief engineer.

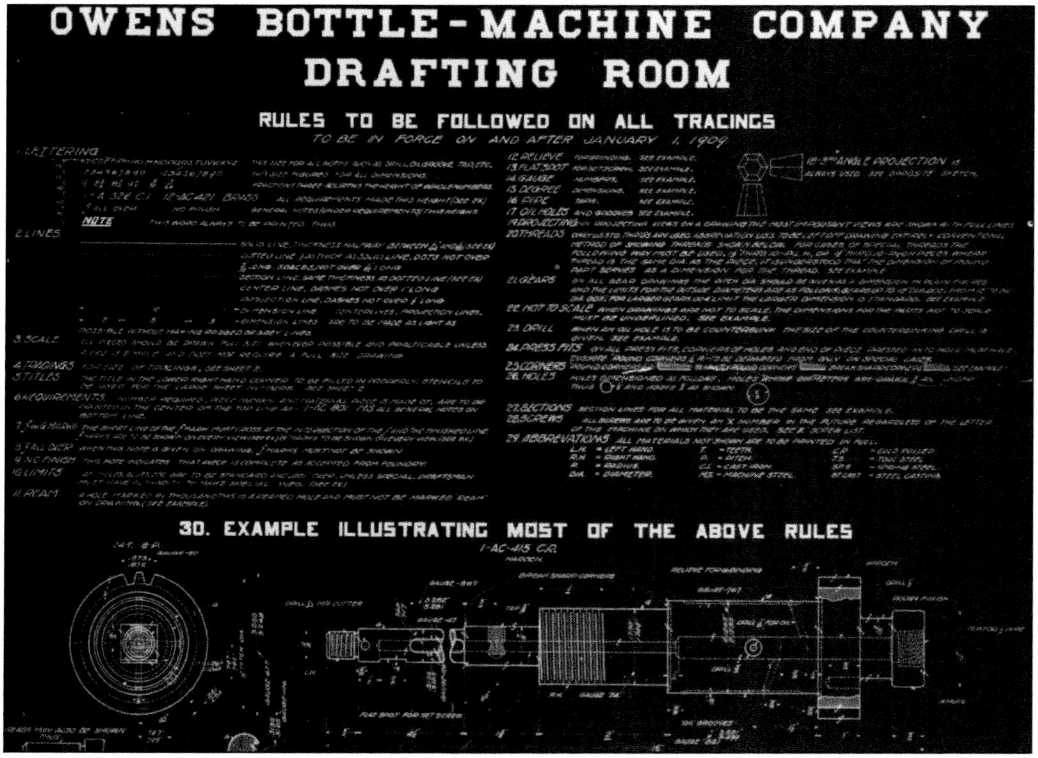

Bock and LaFrance took machine building to a fine art. Standardized parts, tools, jigs, and gauges were pioneered in Owens Bottle even before Henry Ford or Alfred Sloan at General Motors. LaFrance's rule allowed for the building of identical machines worldwide with interchangeable parts. (Courtesy of the University of Toledo's Ward Canaday Center.)

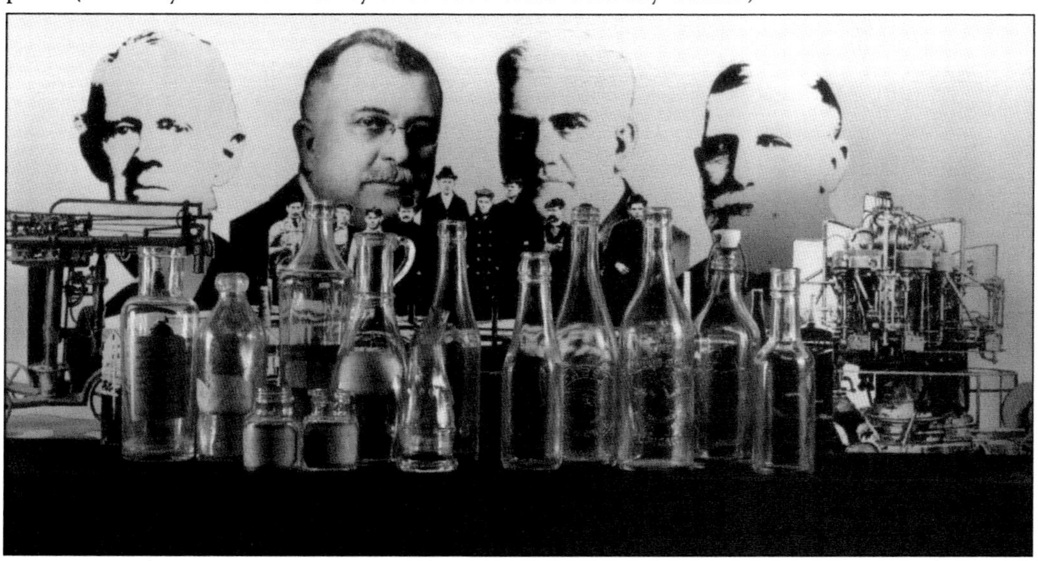

Pictured are Edward Libbey (second from left) and Owens (second from right) with the Graham brothers on each side. The Graham brothers had developed a type of automated bottle feeding system, which Owens bought the patent rights to in 1907 and incorporated in later machine designs.

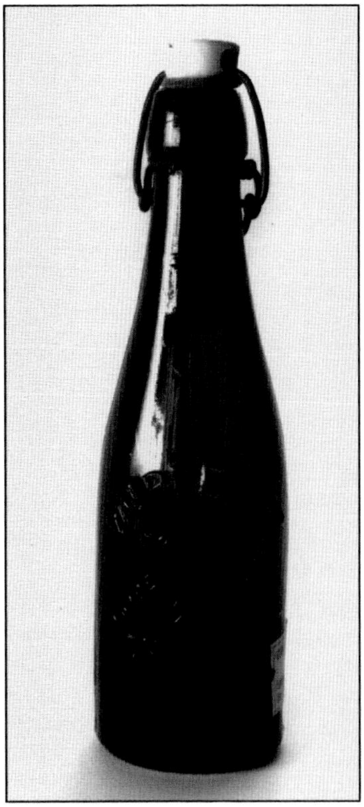

Prior to Michael Owens's automatic blowing machines, there were no beer bottles, milk bottles, or baby bottles. Bottles were expensive and used in a limited amount. Whiskey bottles and drug bottles made up most of the market. (Courtesy of the University of Toledo's Ward Canaday Center.)

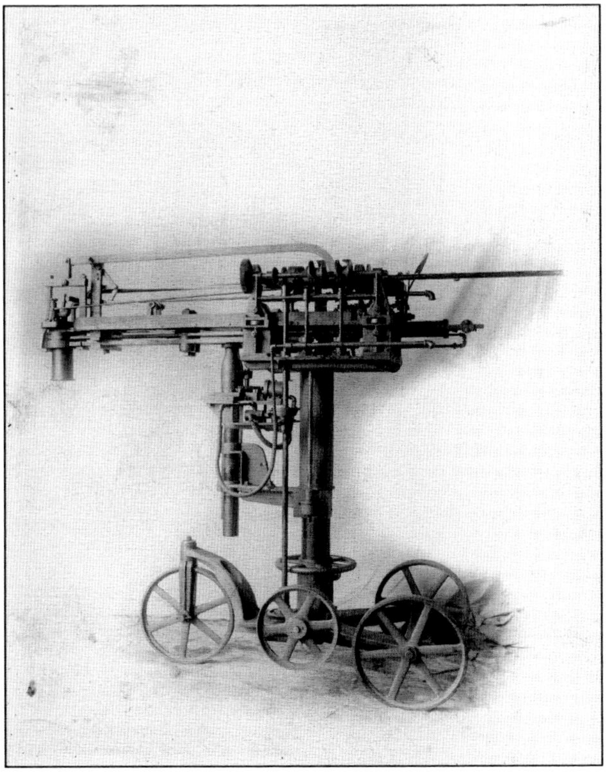

Owens's Machine Two was a mounted bicycle pump in 1900. The machine successfully produced bottles but was not considered fully automated. (Courtesy of the Toledo Lucas County Library.)

This was the first real commercial machine. The improved engineering reflects the influence of Emil Bock on the team. This machine was manufactured from 1904 to 1908. (Courtesy of the Toledo Lucas County Library.)

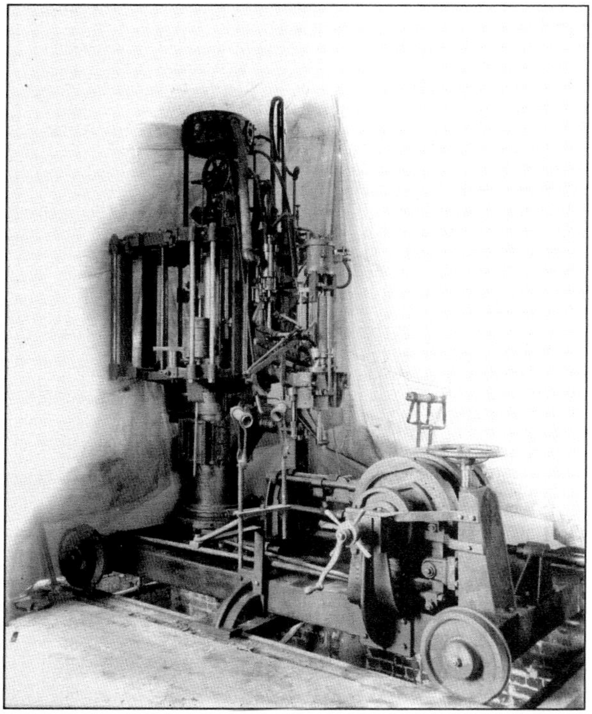

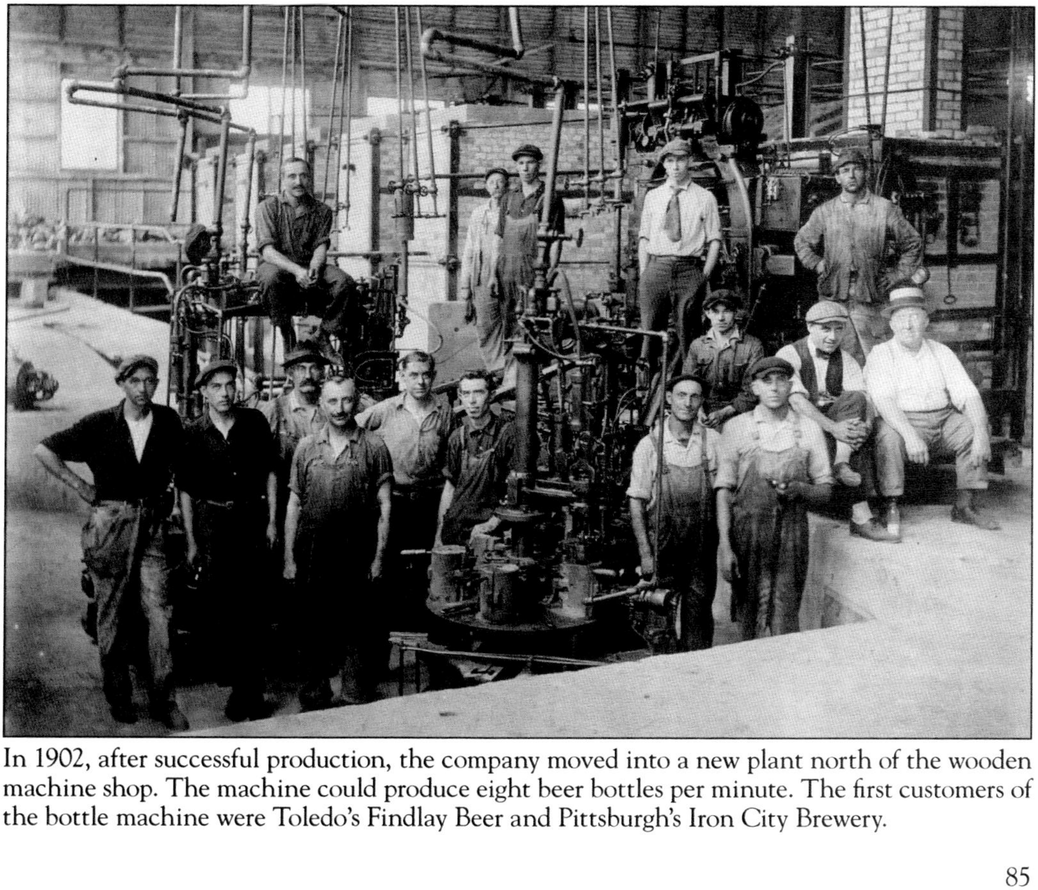

In 1902, after successful production, the company moved into a new plant north of the wooden machine shop. The machine could produce eight beer bottles per minute. The first customers of the bottle machine were Toledo's Findlay Beer and Pittsburgh's Iron City Brewery.

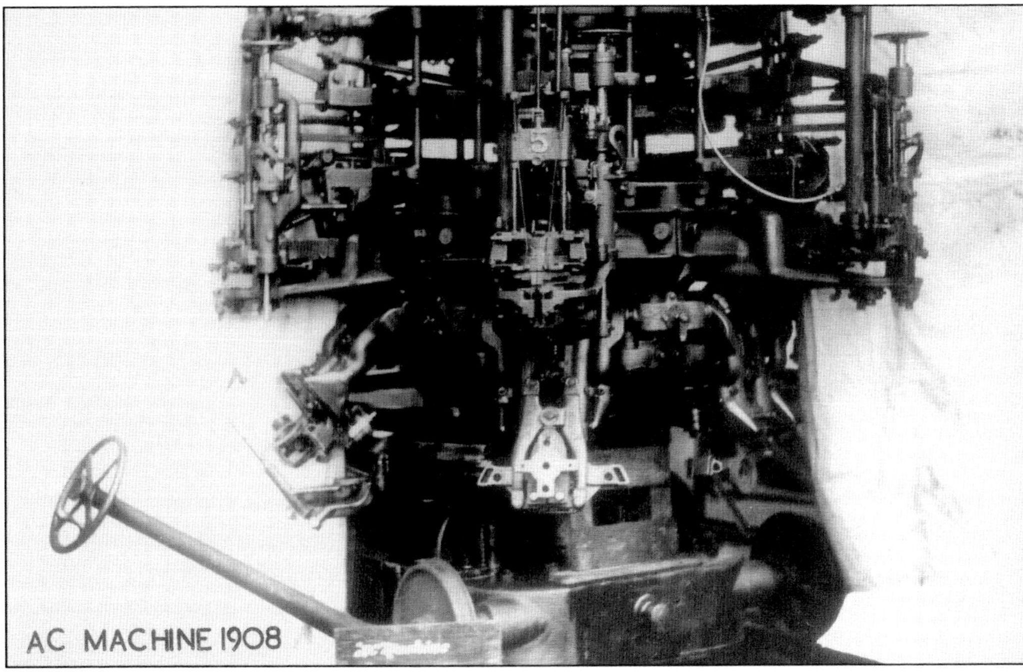

The AC machine 1908 could produce 150 gross of pint beer bottles per day. It became a workhorse of several Pittsburgh brewers and was one of the first machines manufactured for leasing in Europe. The major improvement over the A machine was less bottle distortion. (Courtesy of the University of Toledo's Ward Canaday Center.)

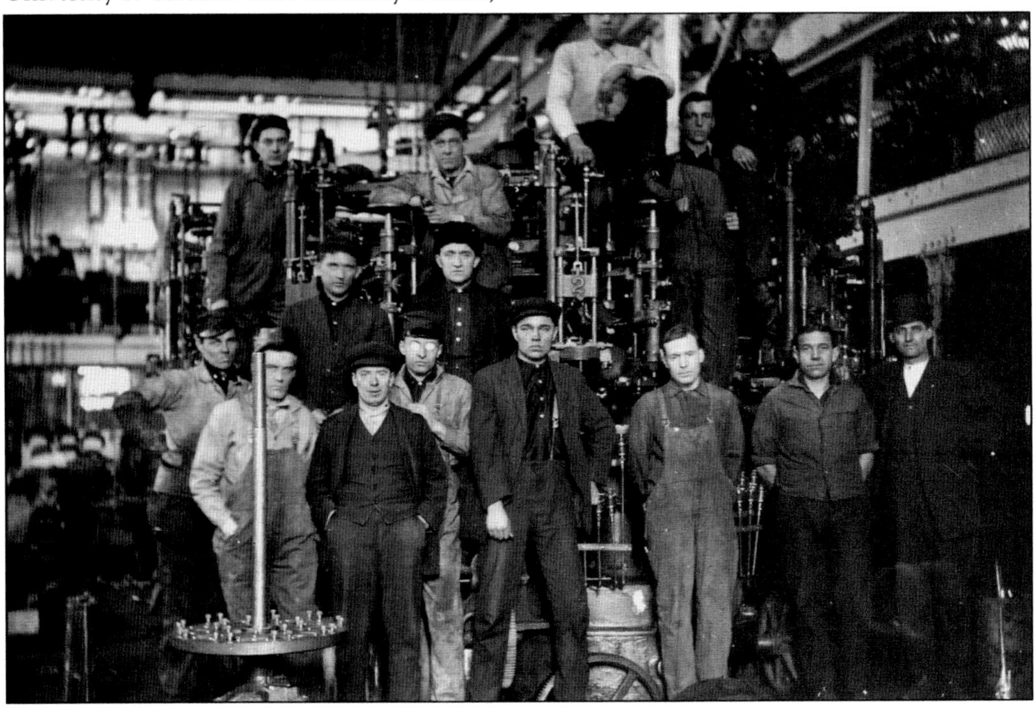

The crew here is assembling an AD machine for H. J. Heinz. The machine would make 24 ketchup bottles a minute.

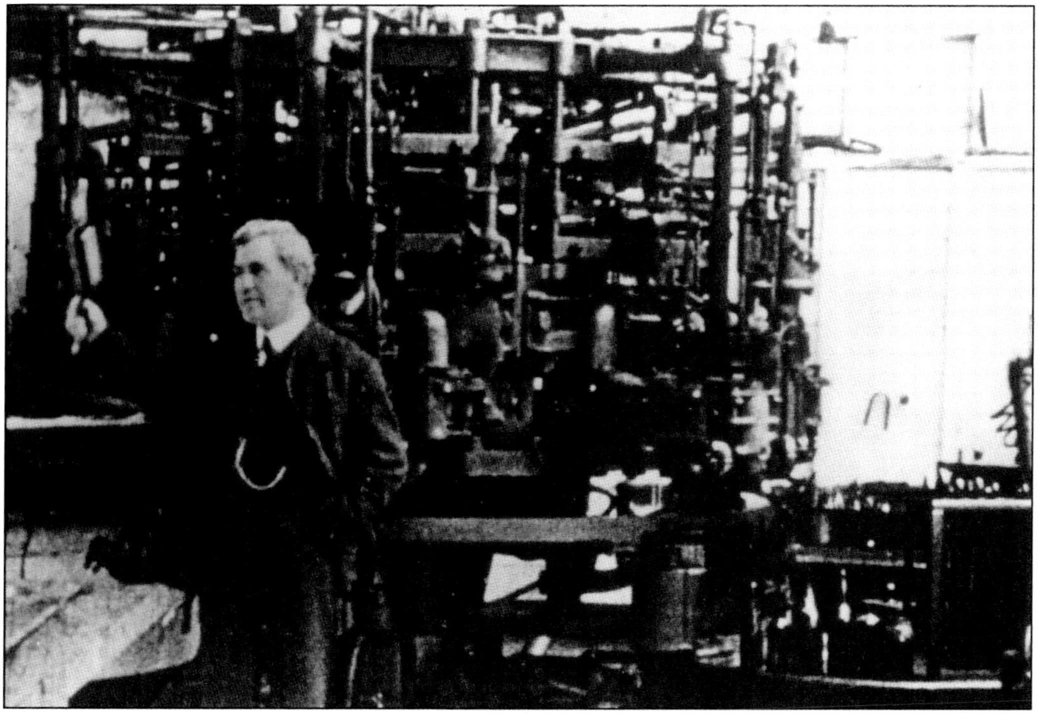

This picture has appeared often and was taken from a 1910 industrial film of the automatic bottle-making machine.

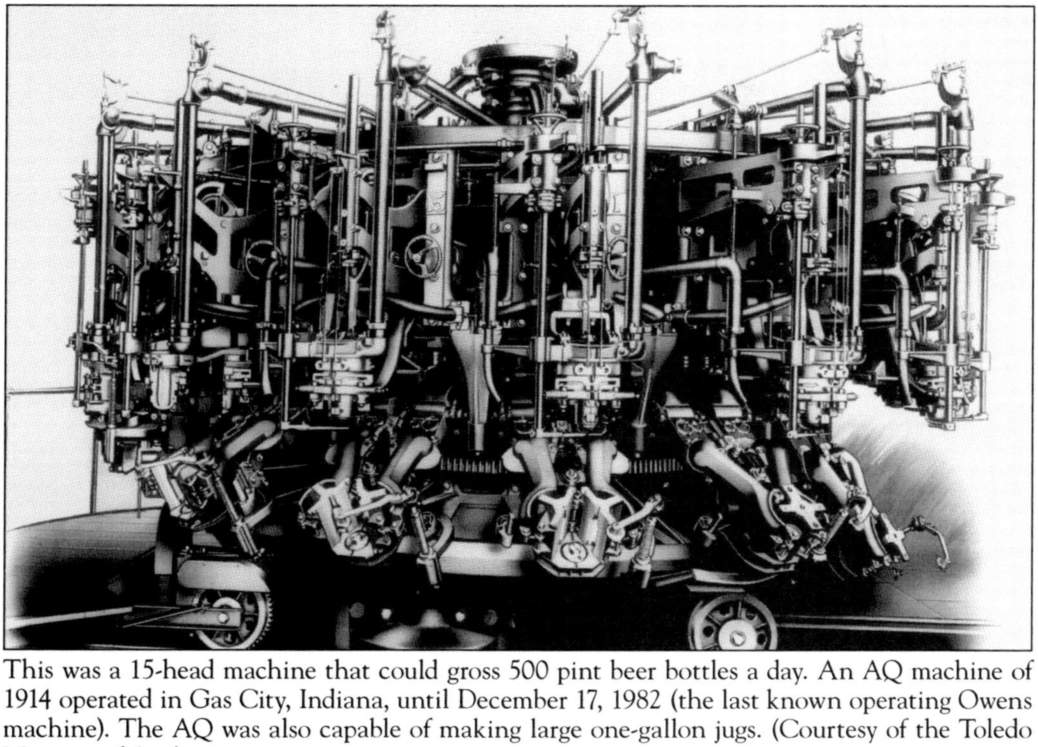

This was a 15-head machine that could gross 500 pint beer bottles a day. An AQ machine of 1914 operated in Gas City, Indiana, until December 17, 1982 (the last known operating Owens machine). The AQ was also capable of making large one-gallon jugs. (Courtesy of the Toledo Museum of Art.)

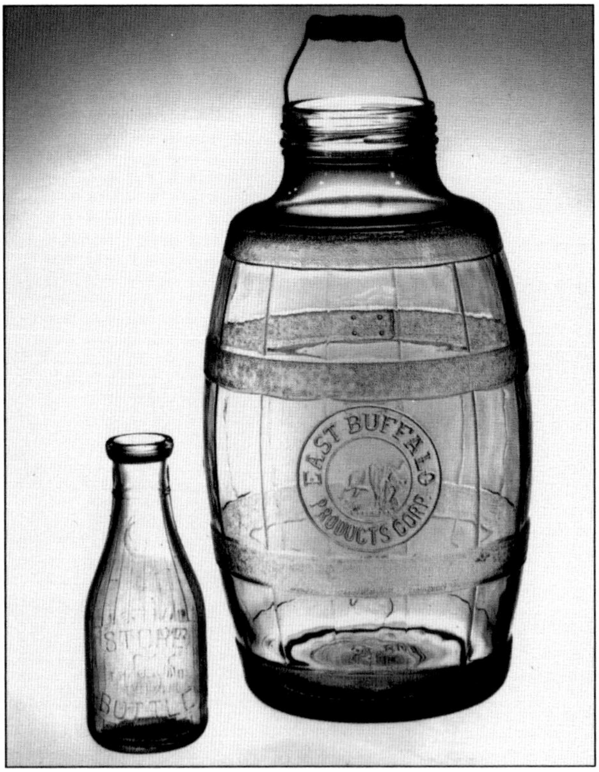

The AQ machine produced larger bottles and jars as well as pint beer bottles. The Mason screw-type jar was actually invented in 1859 as iron molds came into use. (Courtesy of the University of Toledo's Ward Canaday Center.)

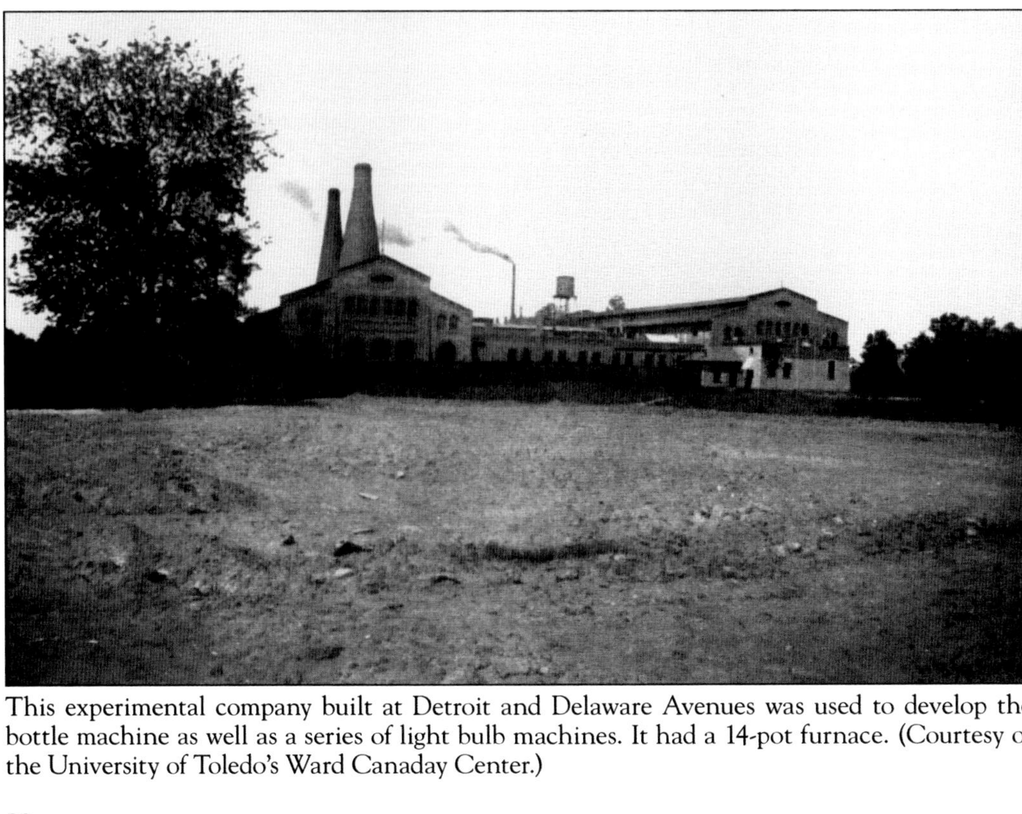

This experimental company built at Detroit and Delaware Avenues was used to develop the bottle machine as well as a series of light bulb machines. It had a 14-pot furnace. (Courtesy of the University of Toledo's Ward Canaday Center.)

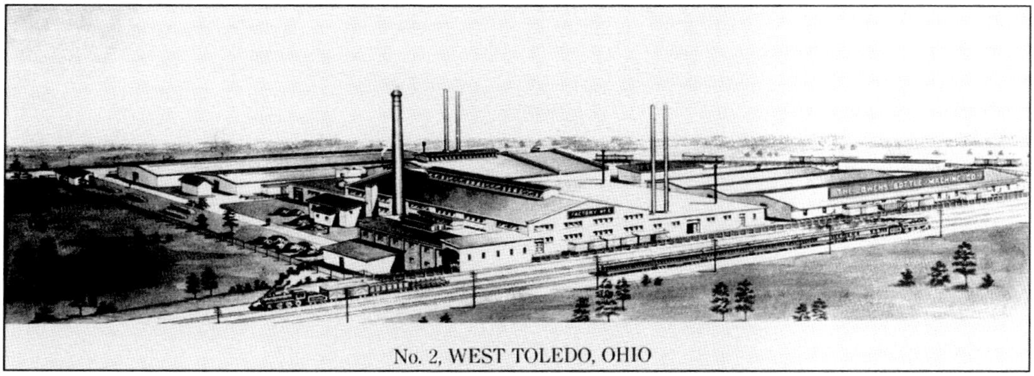

The West Toledo plant was originally built as a demonstration plant. For years, however, the plant produced wine and brandy bottles. Mike Owens's brother worked as a supervisor at this plant. (Courtesy of the University of Toledo's Ward Canaday Center.)

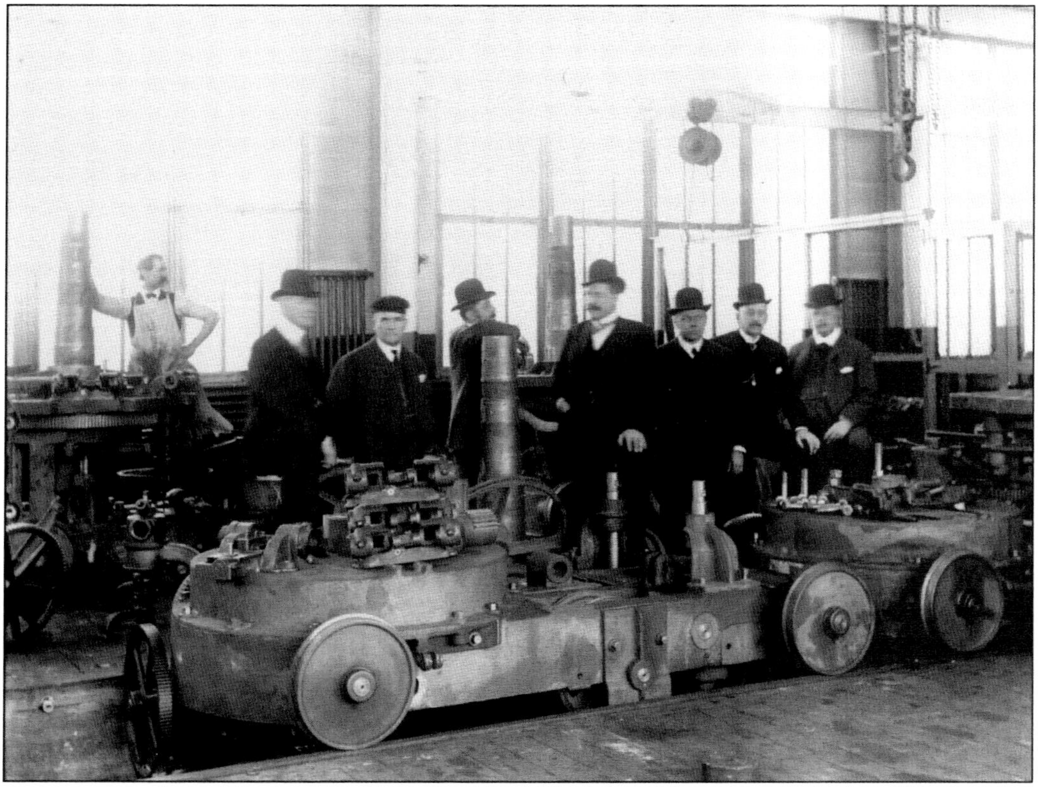

Owens and Edward Libbey's strategy was to license the right to use the bottle machine around the world, instead of selling the machines directly. As early as 1906, they formed a company to manufacture bottle machines for leasing in Manchester, England. Mike Owens is shown here third from left, supervising the construction of bottle machines.

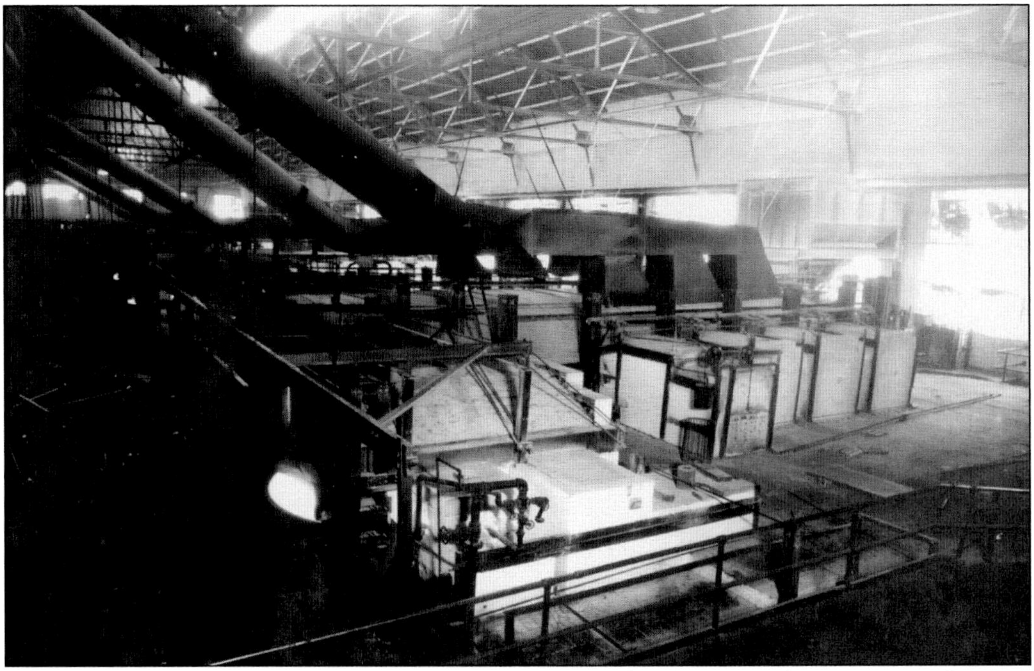

One of the necessary steps to commercialization was the development of a continuous furnace. Molten hot glass needed to be supplied to assure the high production rates of Mike Owens's later machines.

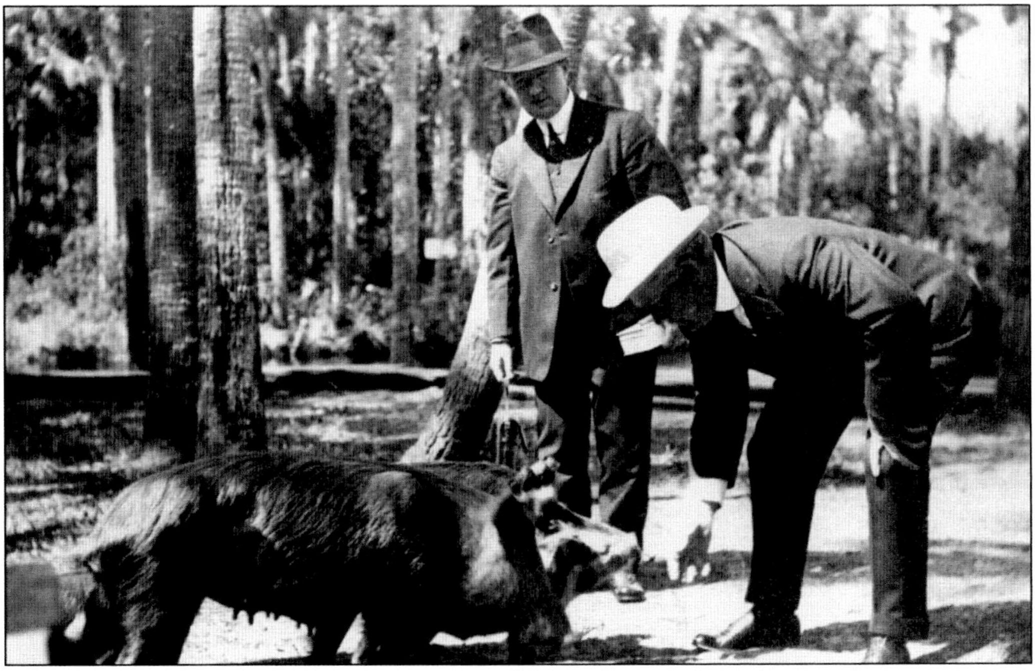

Owens worked closely to Tom Rowe, president of the American Flint Glass Workers, to minimize union job loss with the use of automated bottle machines. Owens had been a union officer in West Virginia before joining Libbey Glass. Both men were close friends and golfing buddies. To their credit, jobs and wages actually increased with the spread of automated bottle machines.

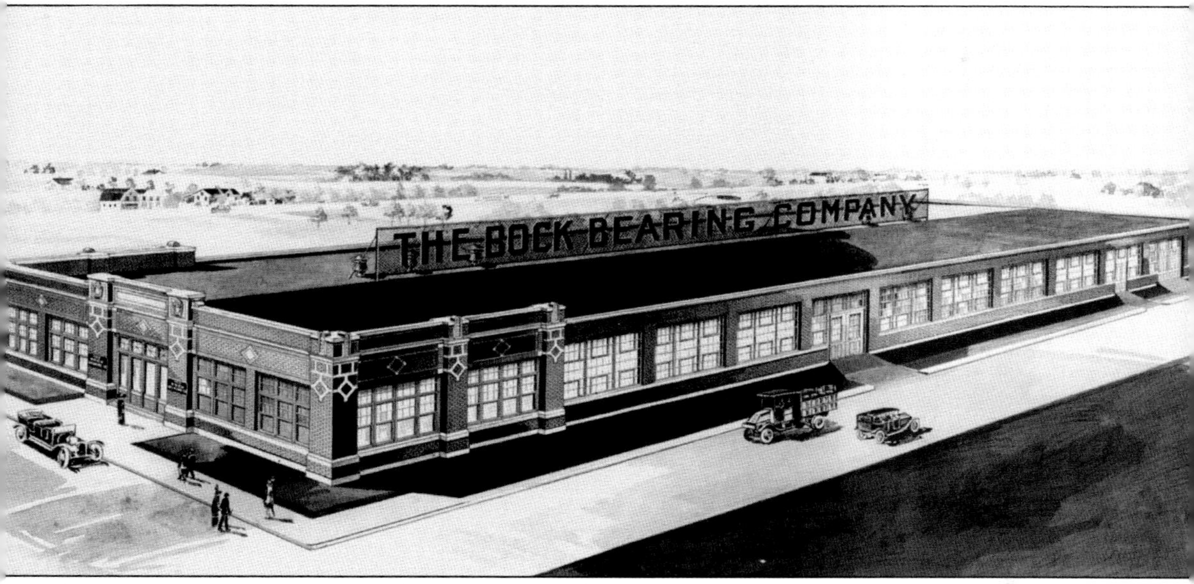

Emil Bock resigned from Toledo Glass in 1910 but went on to found a number of companies. He patented the now-famous tapered roller bearing and built a Toledo plant for its manufacture. (Courtesy of the University of Toledo's Ward Canaday Center.)

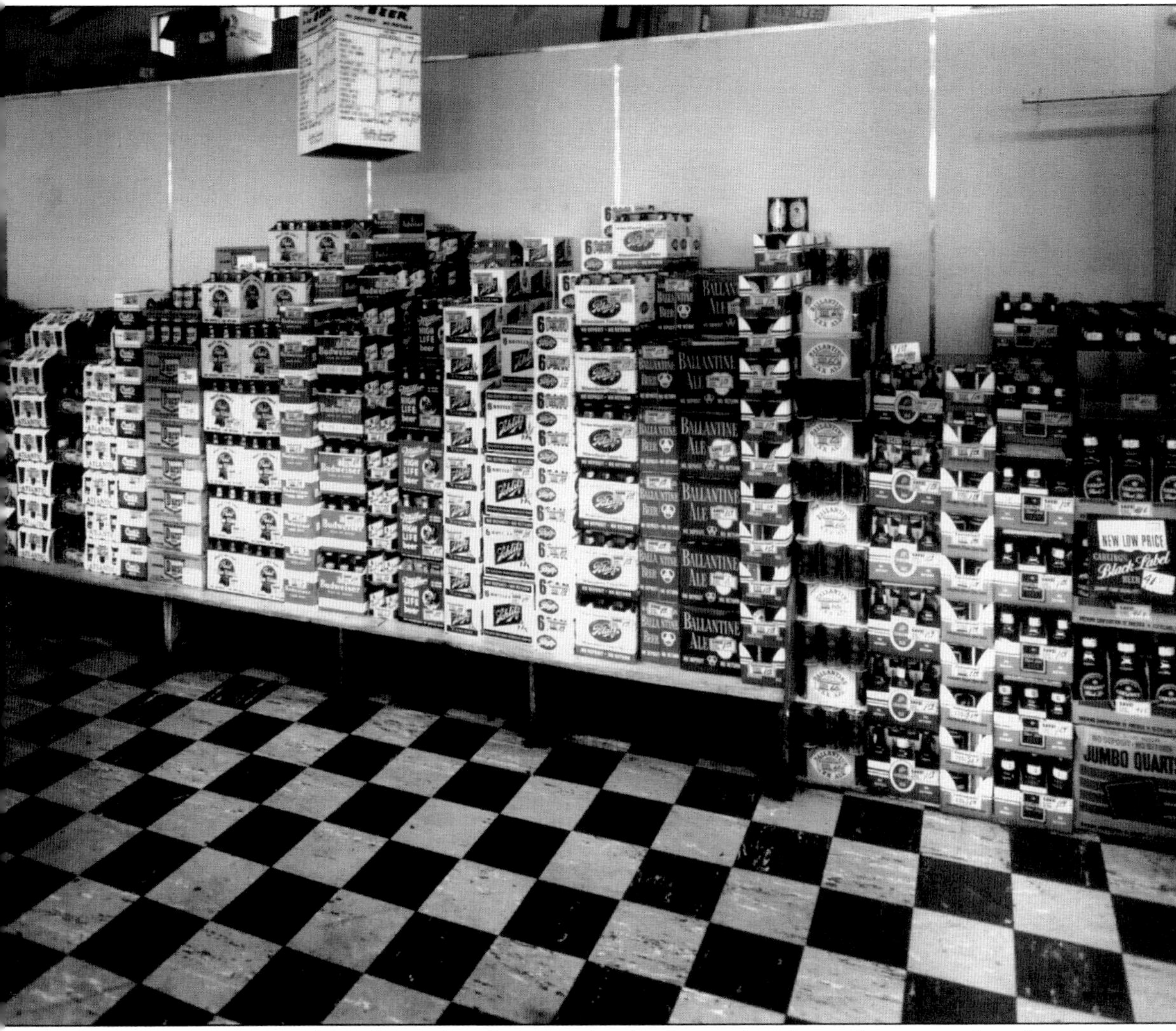

The first commercial bottle produced in 1904 by the Owens bottle machine was a beer bottle for Toledo's brewery Finday Brewing. By 1950, over 90 percent of beer bottles were manufactured on Owens-type machines. Rates of over 300 bottles a minute were achievable by then.

Eight

The Plate Glass Revolution

Flat window and plate glass represented a unique segment of the glass market in 1910. The thin and small piece window glass was produced by the blown cylinder method versus the colonial crown or bull's-eye method formed by spinning a gob of glass (known as a parison) on the end of a rod. The thicker plate glass was cast on a slab and polished. The window glass industry was ripe to automation because of its labor-intensive processes. Cylinders had to be blown, cooled and cut, flattened, annealed, and cut to size. The American Window Glass Workers limited the number of cylinders that could be produced in a day, further driving the price up. The market pressure was building, however, for thicker and larger windows.

In the early 1900s, Pennsylvanian inventor Irving Colburn was experimenting with an automated flat glass–making machine. Colburn's machine pulled molten glass from the furnace as it cooled into flat glass. In 1909, Colburn had an experimental flat glass plant in western Pennsylvania. A year later, Michael Owens toured this experimental plant looking for ways to automate flat glass. When Colburn Company failed and he could not raise any more capital, Owens brought him to Toledo. Libbey and Toledo Glass started a new developmental program under Owens.

Toledo Glass poured in its resources, including Owens and engineer Richard LaFrance's time, into the developmental work on the Colburn machine known as CX. Success came on Thanksgiving Day, 1913, with the run of 200 feet of flat glass. The machine pulled the glass sheet from a molten bath and flattened it for cooling. Libbey formed a new company, Libbey-Owens Sheet Glass. Owens did more than aid in the implementation of the Colburn continuous machine, he designed a fully automated glassmaking facility in West Virginia to fully realize its potential. The plant used conveyor belts and materials handling equipment. Owens did for continuous materials production what Henry Ford did for assembly line manufacture. Owens also improved the glassmaking process itself to further improve the Colburn machine operation. Again Owens proved that automation would result in increased employment and wages in the long run due to market growth.

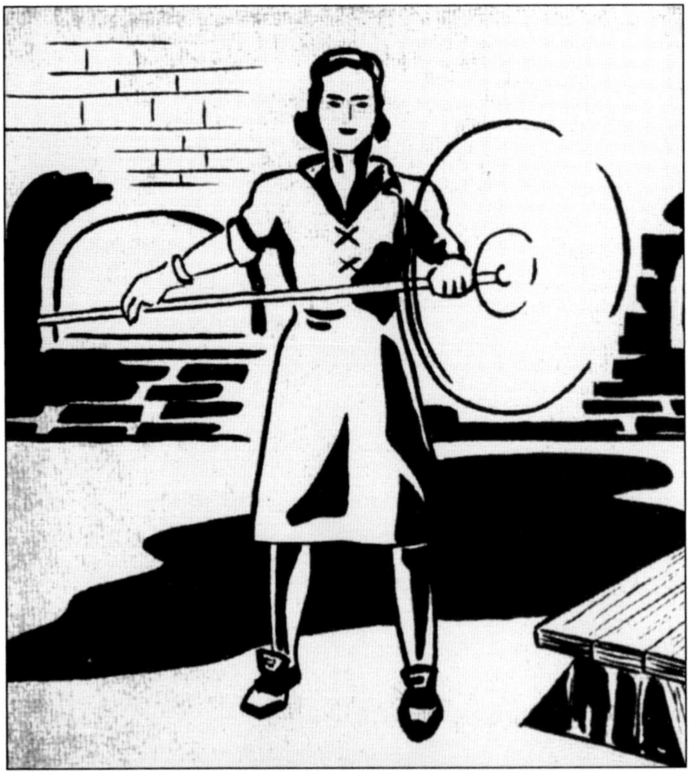

A glassblower would blow a large globe on his blowpipe; the globe was transferred to a pontil rod of an assistant. Then the rod was spun and swirled, producing a flat disk. The disk was then cut into window squares. The pontil left a center mark known as a bull's-eye that could be seen in Colonial glass windows.

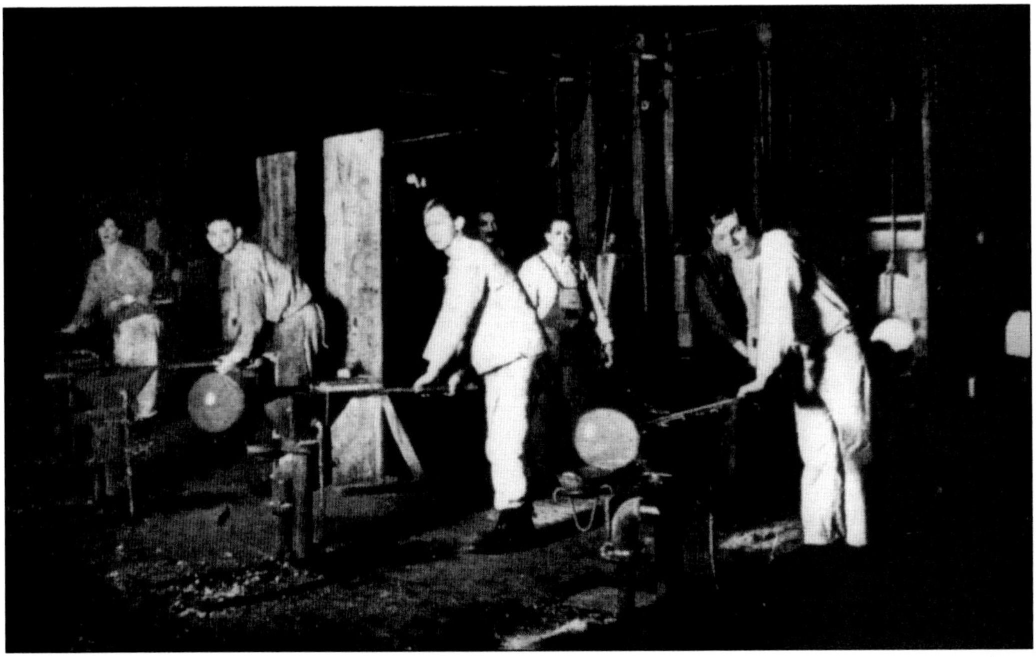

The first step in window making was for a gatherer to form a gob of molten glass. These gathers were Belgian; Belgium is where the craft of gathering and blowing cylinders was perfected. The Maumee Company on Sophia Street was unique in that it was fueled by coal from 1902 to 1928.

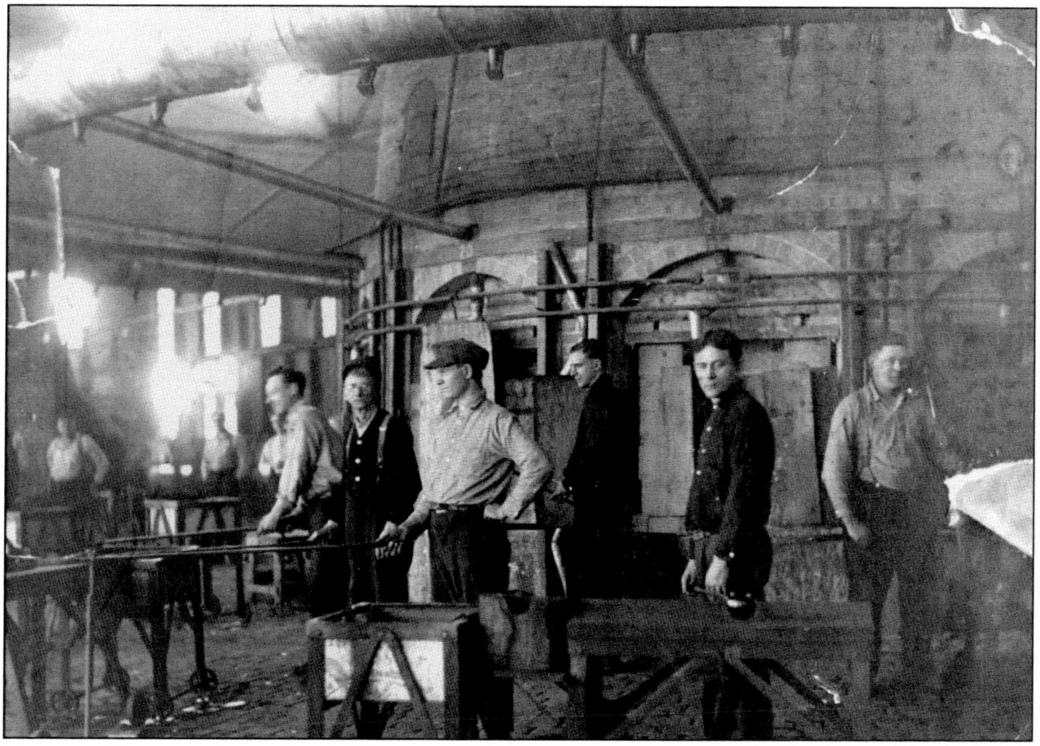

Maumee Glass was formed in 1889 to produce window glass from blown cylinders.

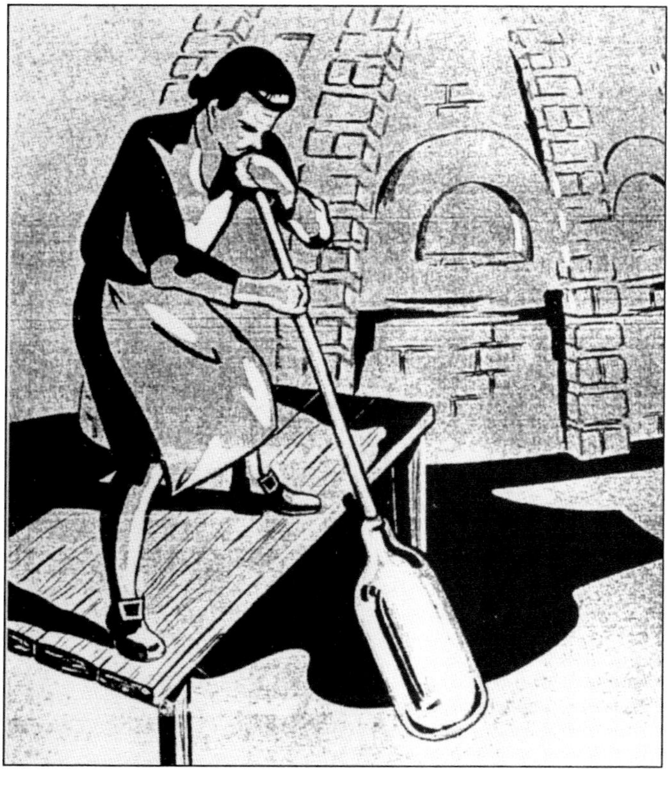

Cylinder glass goes back at least to Colonial times. Baron Stiegel was producing cylinder glass prior to the American Revolution in Pennsylvania. The older competing process was known as crown glass with its distinctive bull's-eye.

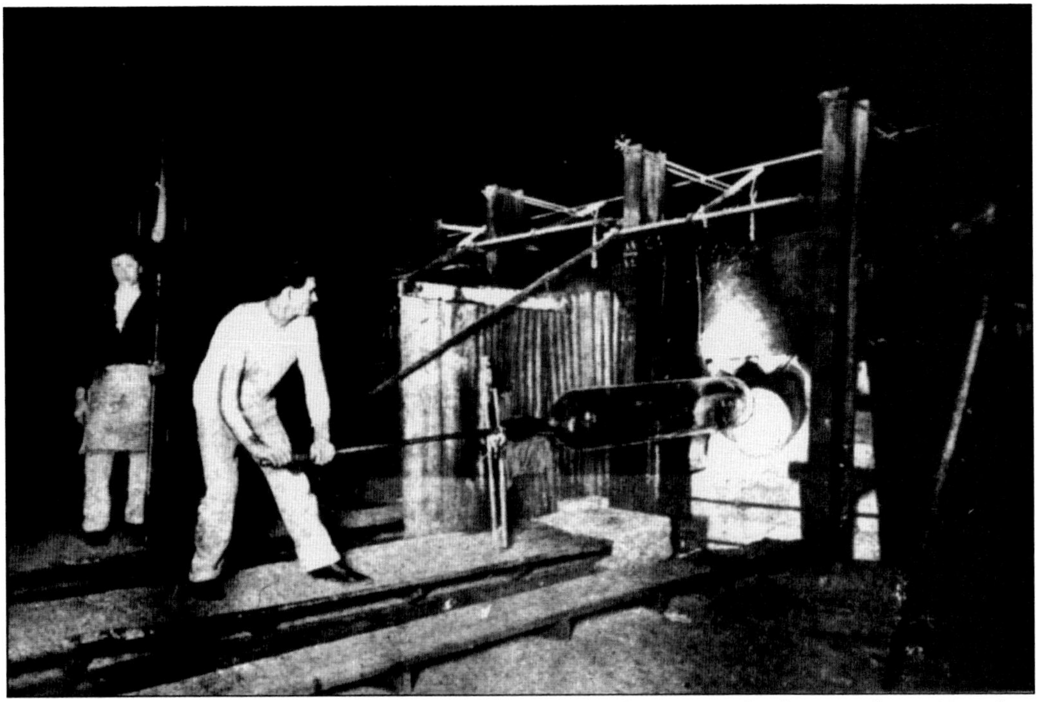

Cylinder glassblowing took skill as well as a set of strong lungs. A cylinder was about 40 inches by 85 inches. A skilled Belgian cylinder blower could make 120 cylinders a day, but in American plants the union restricted output to 72 cylinders per man per day.

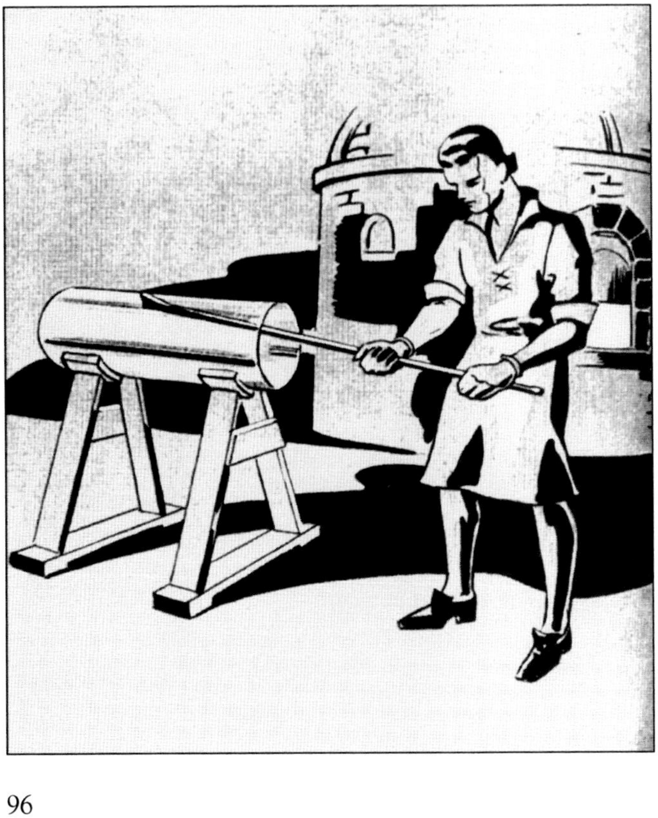

Once a blown cylinder had its ends or caps cut off, the "snapper" took over. The typical cylinder was 12 to 40 inches in diameter and about 70 inches to 80 inches long. The cylinder tube was split lengthwise by running a hot iron along the inside and then applying a cold iron, causing the cylinder to snap.

The snap cylinder was then heated and flattened. The flattened glass did have a slight distortion in it. Cylinder window glass was commonly used in middle-class homes, while distortion-free polished plate was used for windows of the wealthy. Over 30 window glass companies operated in northwest Ohio during the years 1888 to 1917. In 1917, the Owens/Colburn sheet glass machine replaced the cylinder glass method.

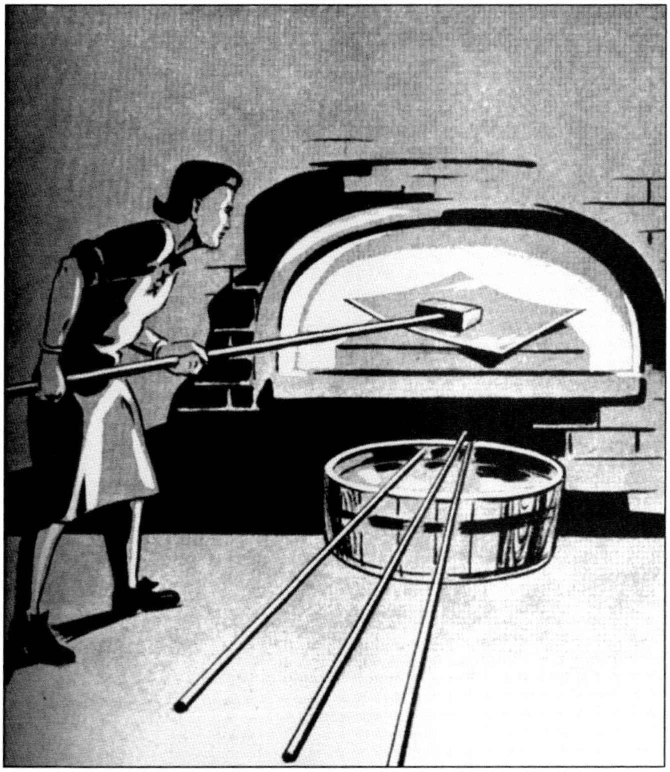

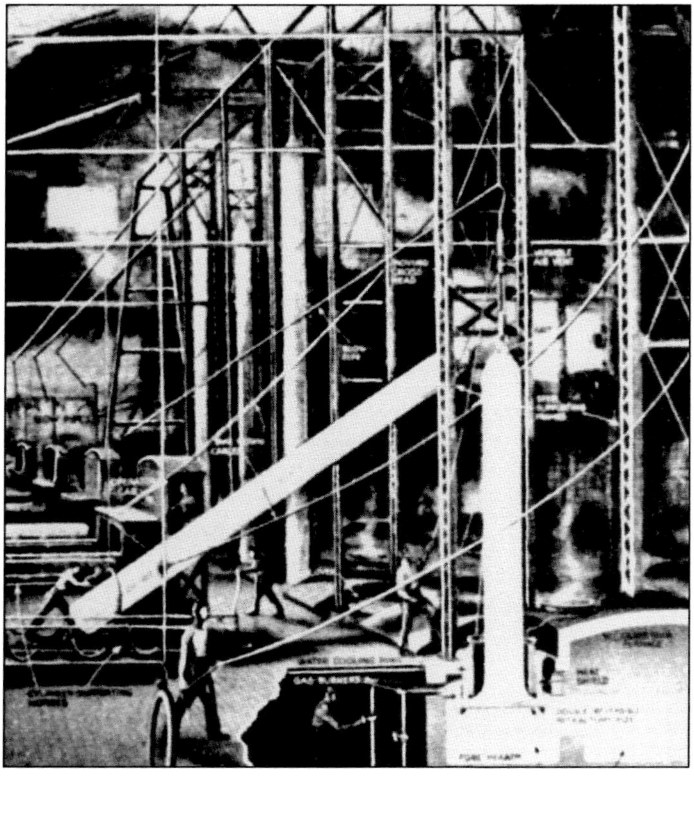

The automated Lubbers system for blown cylinders was successfully resisted by the glass unions for 10 years. The mechanized cylinder-blowing process of John Lubbers was patented in 1900. The union resistance, while successful in the short run, would be overwhelmed by the implementation of the Colburn machines of Owens-Libbey Sheet Company.

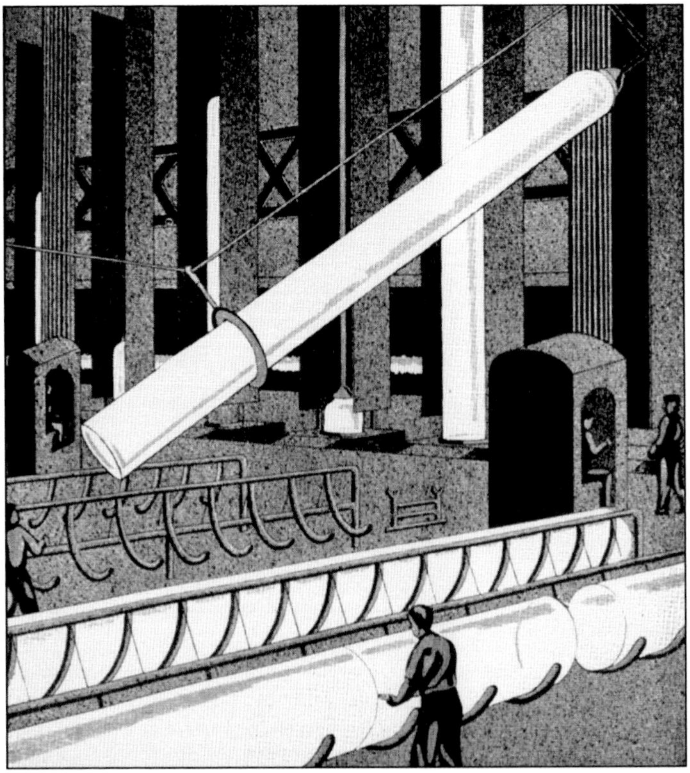

In 1909, Case and Merry (formerly Maumee Cooperative) owned the patent rights to the Lubbers process but lacked the capital needed to install the process. Years later in the 1920s, the Maumee Chamber of Commerce tried to resurrect the Lubbers process and failed plant, but the Colburn/Owens machine made the process uncompetitive.

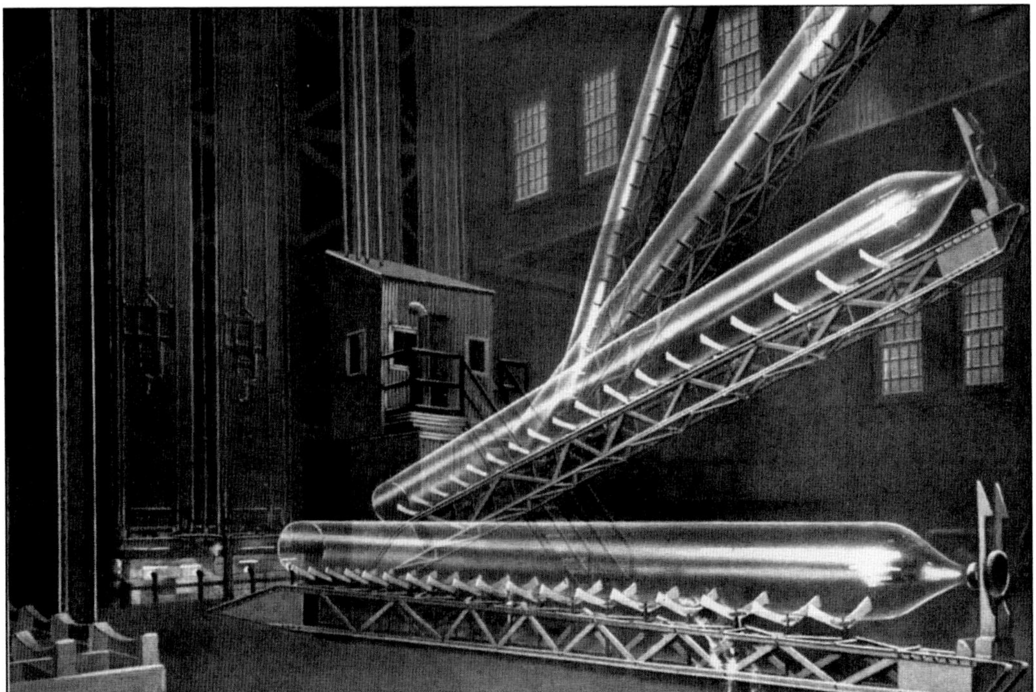

A competing process for the Lubbers process was a similar hot drawing of cylinders. This process was being used by Pittsburgh Plate Glass.

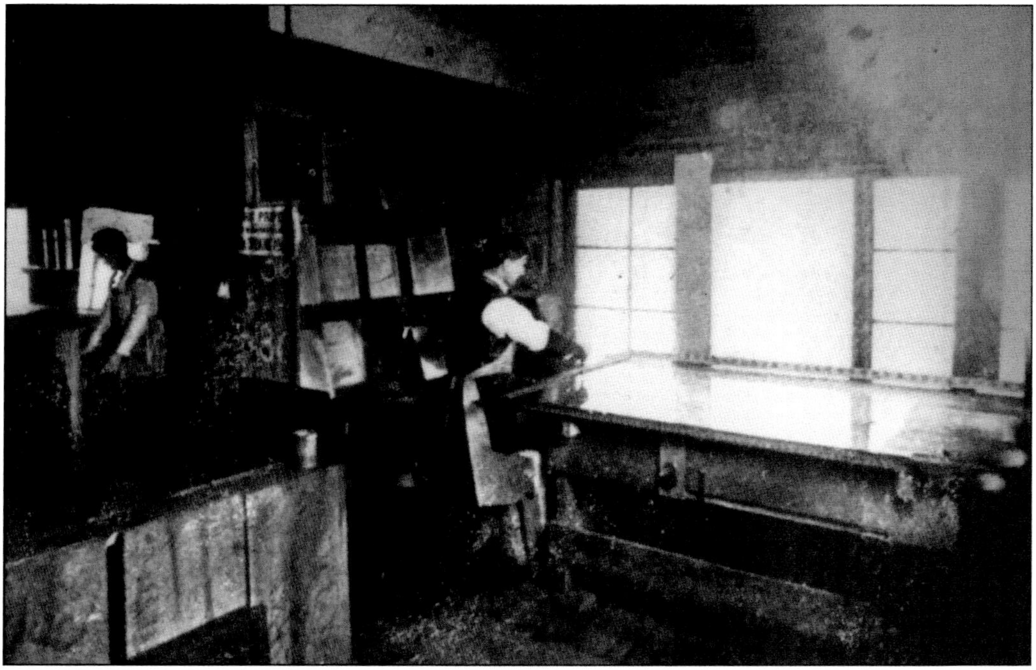

Once flattened, the glass sheet was cut into window squares. Glass cutting to fit non-standard window sizes of the time created an ancillary industry of glass-cutting companies throughout northwest Ohio. (Courtesy of the Rossford Public Library.)

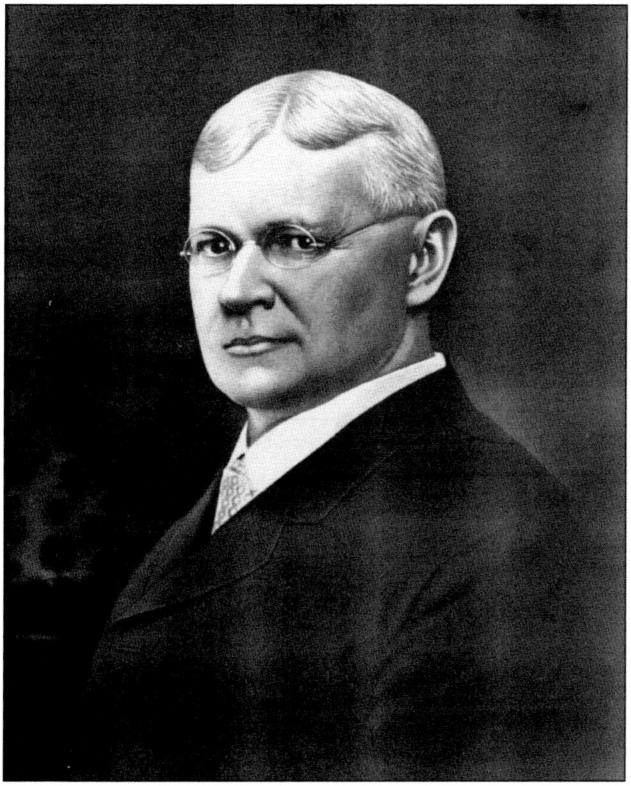

Since 1895, Irving Colburn had worked on the development of a flat glass automated machine. It was not until he teamed up with Michael Owens in 1908 that he was able to fully develop his machine. (Courtesy of the University of Toledo's Ward Canaday Center.)

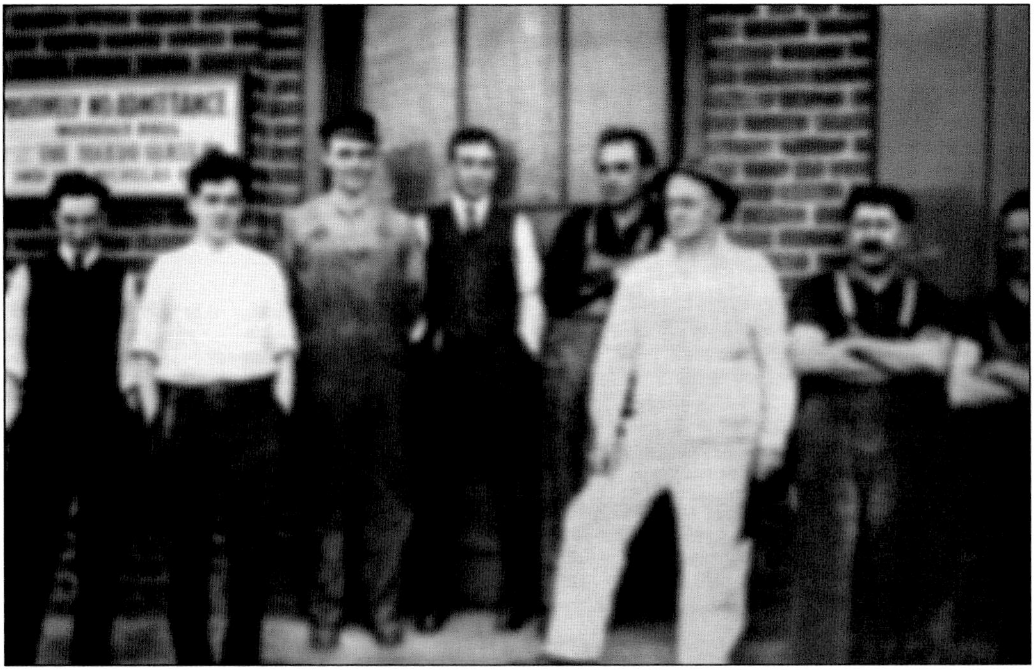

An experimental plant was built on Castle Boulevard for Colburn's development team. Colburn is in white. In May 1913, the team produced the first commercial order of flat glass. A year later, the experimental company became Libbey-Owens Sheet Glass Company. (Courtesy of the University of Toledo's Ward Canaday Center.)

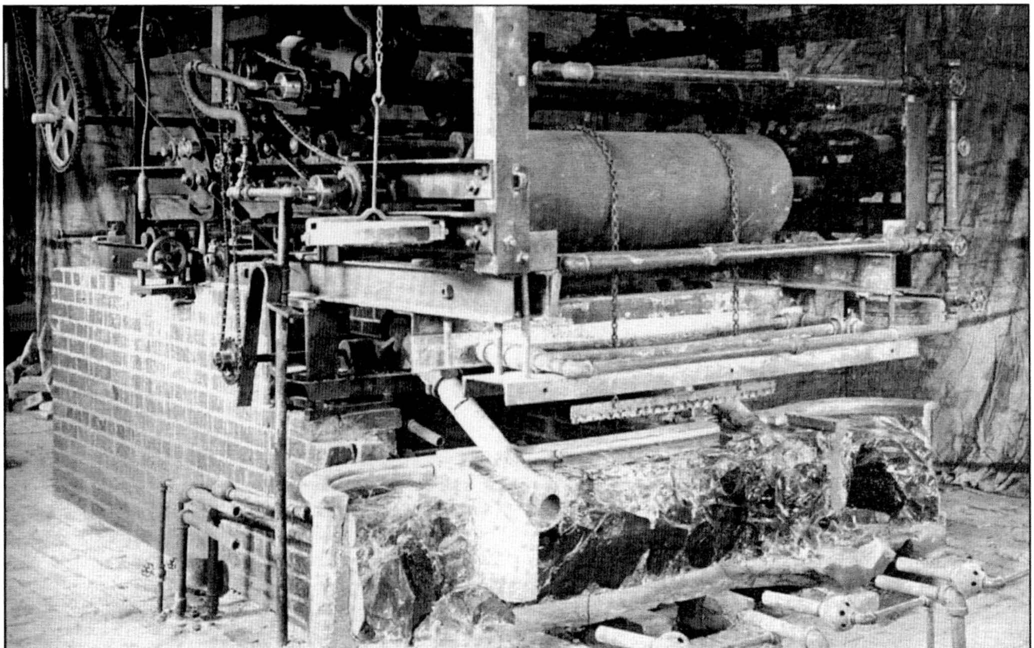

The early window glass machines of Colburn were semiautomatic. His first machine operated at Star Glass in Reynoldsville, Pennsylvania. The machine could produce a piece of window glass that was 16 feet wide by 90 feet long. (Courtesy of the Rossford Public Library.)

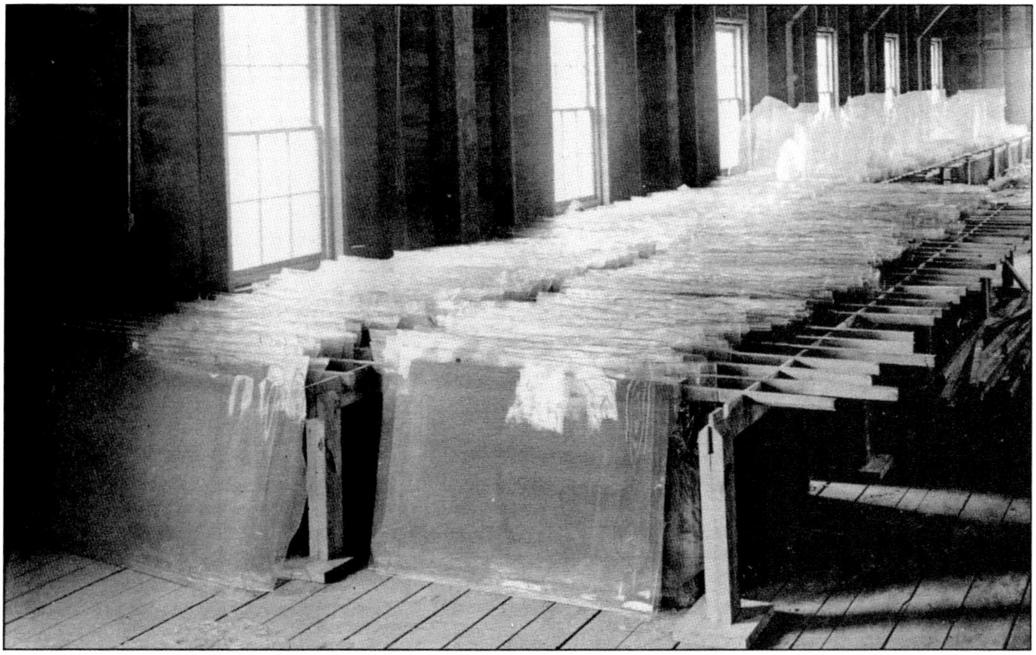

Seen here are early production glass sheets from the Colburn machine from his Reynoldsville, Pennsylvania, experimental plant in 1909. In 1910, Ownes and Libbey bought the rights to the not-fully-developed machinery and brought it to Toledo. (Courtesy of the Rossford Public Library.)

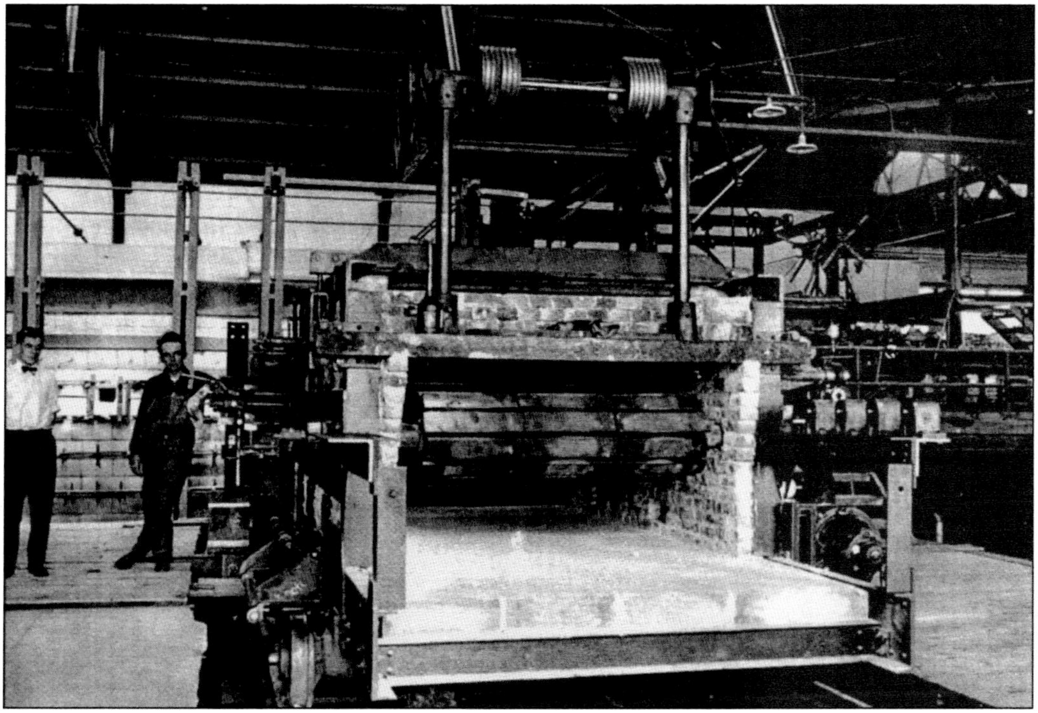

Mike Owens added a chain-link drive to allow for an inline anneal. It was not until 1915 that he had a commercial version.

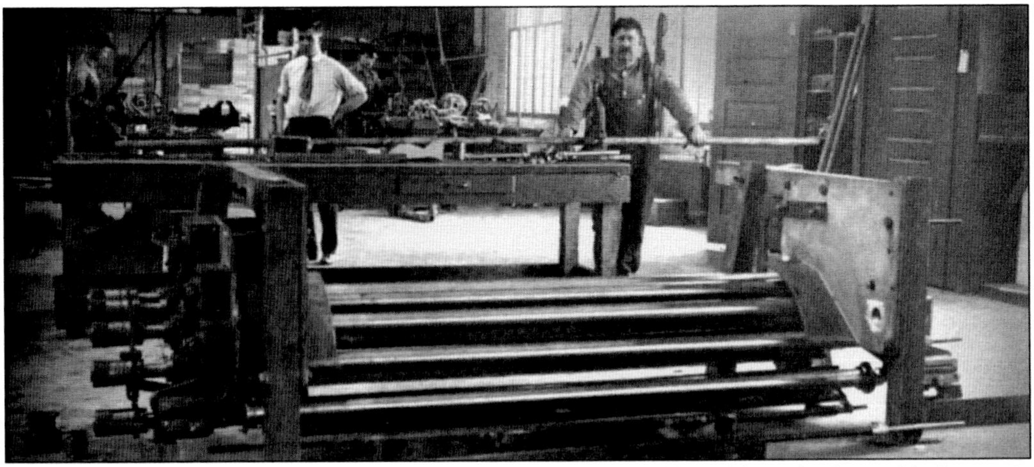

Ford Glass had developed a number of roller convectors and roller tables, which would ultimately be incorporated in the Colburn machine.

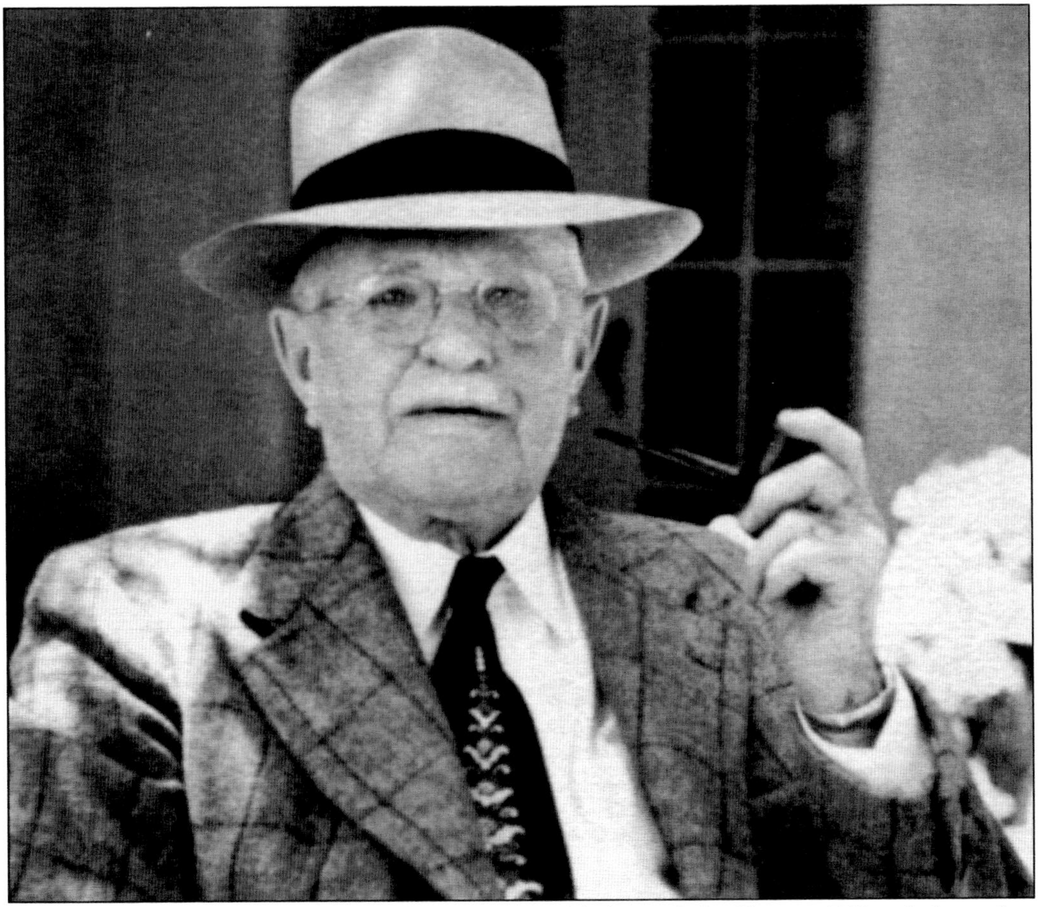

Arthur Fowle, a first cousin of Edward Libbey, was added to Toledo Glass in 1913 to help market Colburn machine–produced glass. He was named to the first board of directors of Libbey-Owens Sheet Glass in 1916. (Courtesy of the Toledo Lucas County Library.)

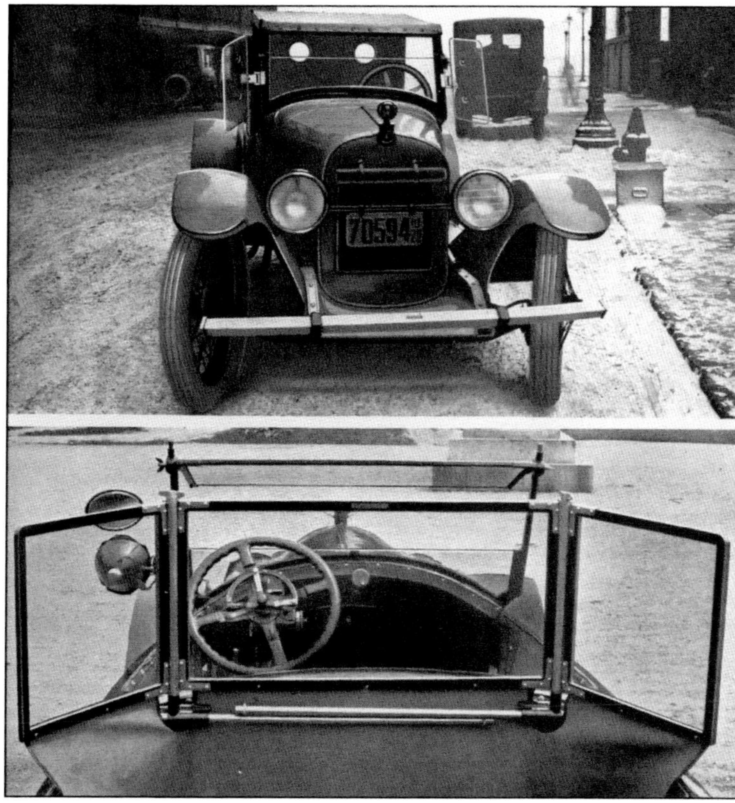

Ford Glass and Pittsburgh Plate glass competed for years for the automotive business. Note that the early glass windows were flat sheets without curvature.

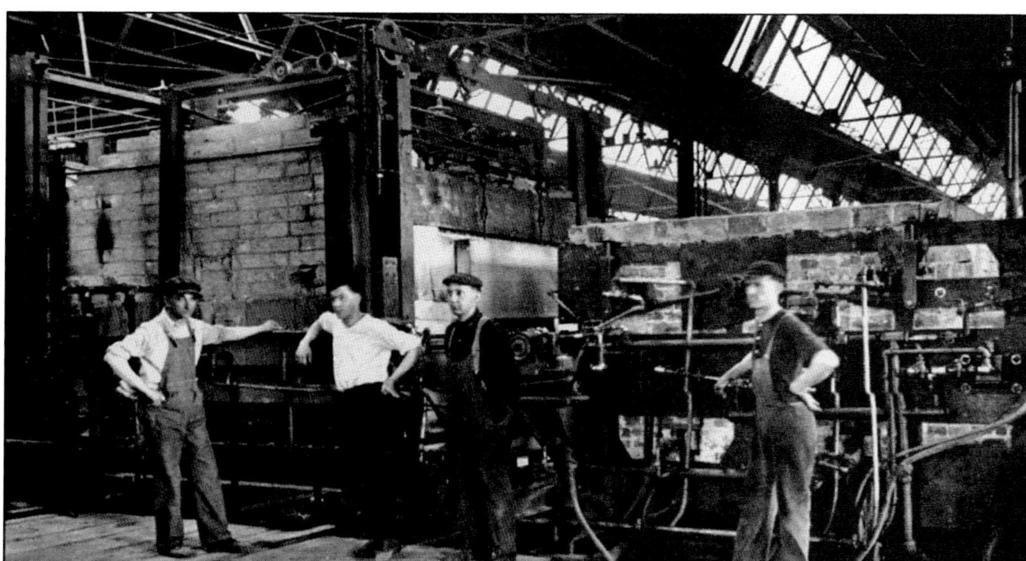

Libbey would build a new company, Libbey-Owens Sheet Glass in 1916 around this new equipment. Libbey-Owens Sheet built its first production plant in West Virginia.

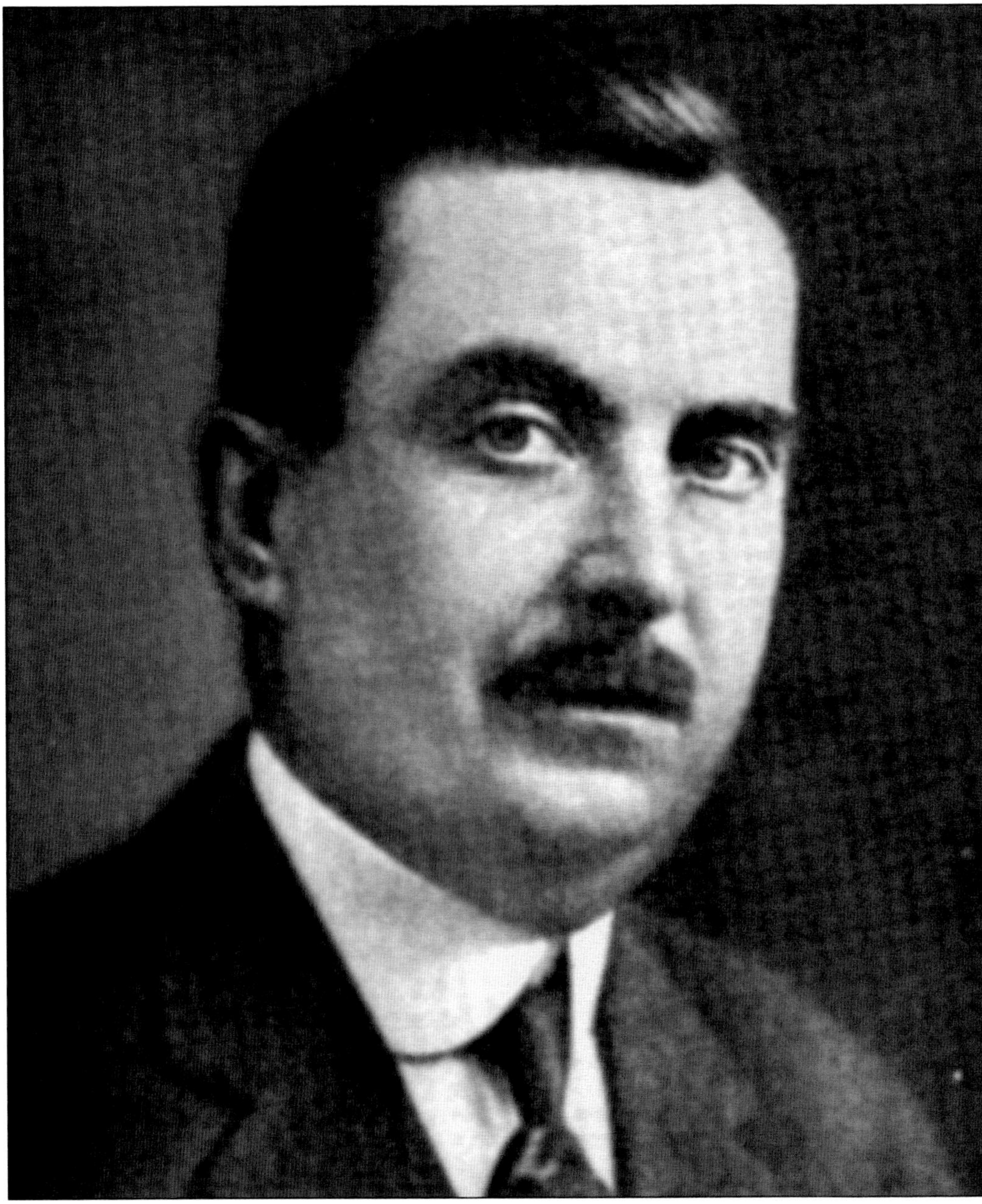

William Boshart was added to the board of directors of the Libbey-Owens Sheet Company in 1915. Boshart was important in selling flat sheet from the Colburn machine. Boshart was also an officer of Owens Bottle and became president of Owens Bottle after the death of Libbey in 1926.

Nine

The Death of the Lions

The automated glass processes of the Toledo faction changed American glass for the next 50 years. In 1999, there were four companies with the Owens Libbey names—Owens-Illinois, Libbey Owens Ford, Owens Corning, and Libbey. Even today there are still Libbey-Owens, Owens-Corning, and Libbey-Owens-Ford.

On December 27, 1923, Mike Owens died while at the Nicholas Building to discuss the production of polished automotive safety glass on the Colburn machines. His pallbearers included two national union leaders—William Clarke and Tom Rowe. It was only after his death that the extent of his generosity became known. He lived by an Irish code of giving that required anonymity with giving.

Owens had been working on his bottle machine, which achieved rates of 800,000 bottles a day at his death. In 1900, prior to the glassmaking revolutions of Owens, a production valued fewer than $20 million, and at Owens death it was over $150 million. Employment in the industry had risen from about 53,000 to 70,000. Wages had increased about $500 a year to $1,300. At his death, child labor in the glass industry had been eliminated and 94 out of every 100 bottles were being made on Owens machines.

Edward Libbey died of pneumonia on November 13, 1925. Both Libbey and Owens had been developing a "second generation" of Toledo men, such as William Boshart and John Biggers. Biggers ultimately rose to vice president of Owens Bottle Company. Boshart took over as president of Owens Bottle after the death of Libbey. Boshart worked to form the merger of Owens Bottle and Illinois Glass to form Owens-Illinois in 1929. Boshart and James Blair would help bring about the merger of Owens Sheet Glass and Ford Plate Glass into Libbey-Owens-Ford in 1929.

Owens-Illinois still produces 50 percent of the world's containers, and Libbey Glass is still a mainstay in glassware. Libbey Glass continues to be a factor in glassware, and Libbey-Owens-Ford continues in flat glass. The industrial lions of Libbey, Owens, Walbridge, and Ford are gone, but the glass focus remains.

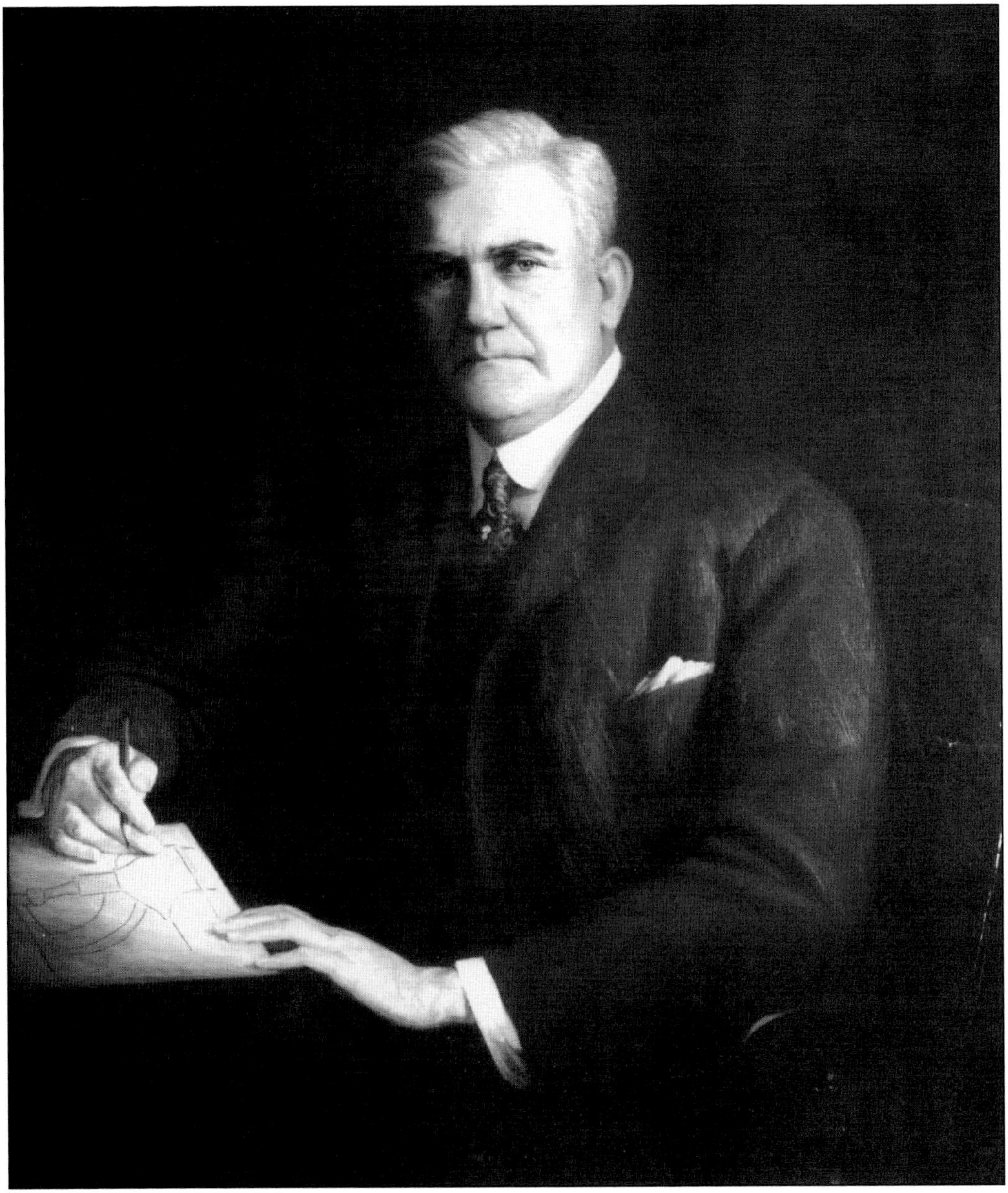

Michael Owens died at a company board meeting. He is shown in this painting with one of his famous hand stretches. Amazingly, Owens could not read a blueprint, but always had paper or a blackboard installed for his rough hand drawings of new inventions. (Courtesy of the University of Toledo's Ward Canaday Center.)

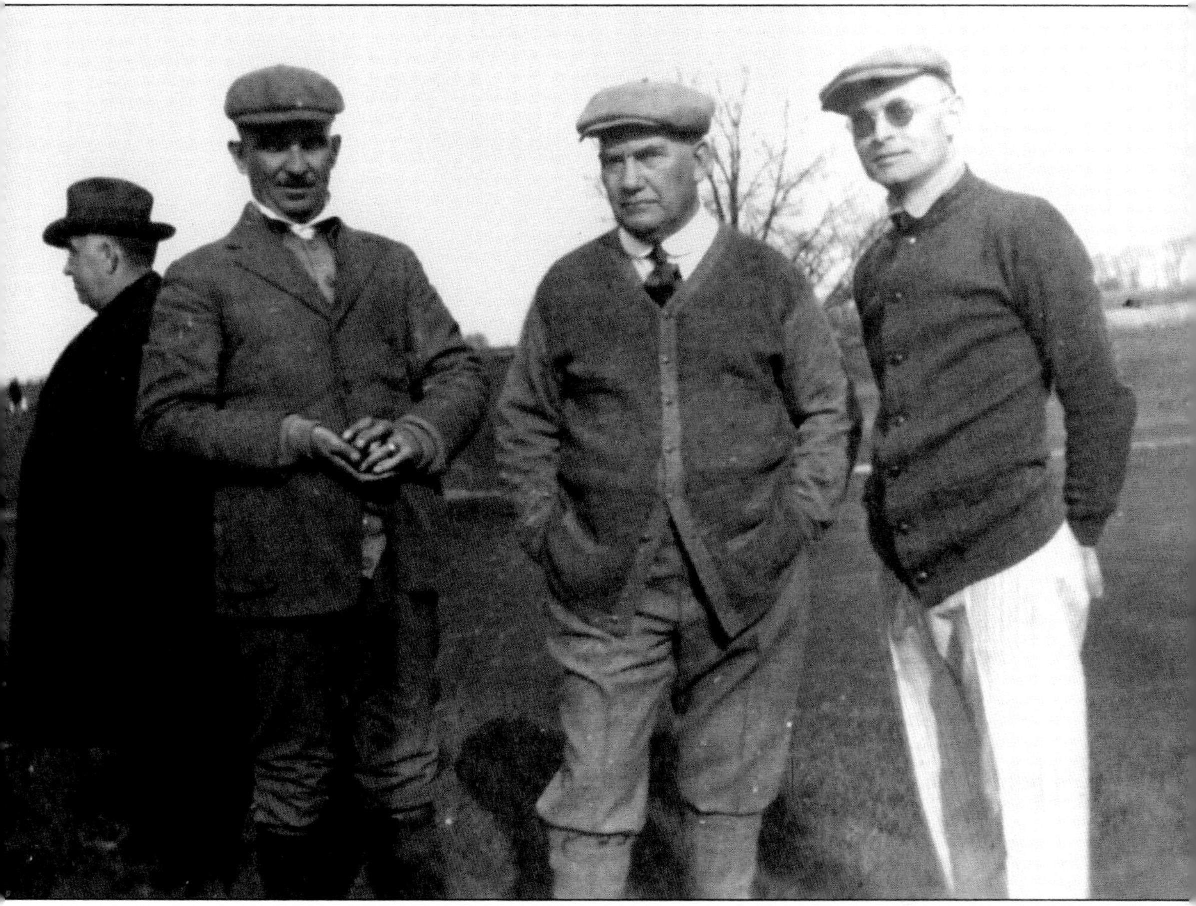

Owens became a passionate golfer and a member of Inverness in the early 20th century. Pictured here are Tom Rowe (right), president of the American Flint Glass Workers, Michael Owens (center), and Irving Colburn. (Courtesy of the University of Toledo's Ward Canaday Center.)

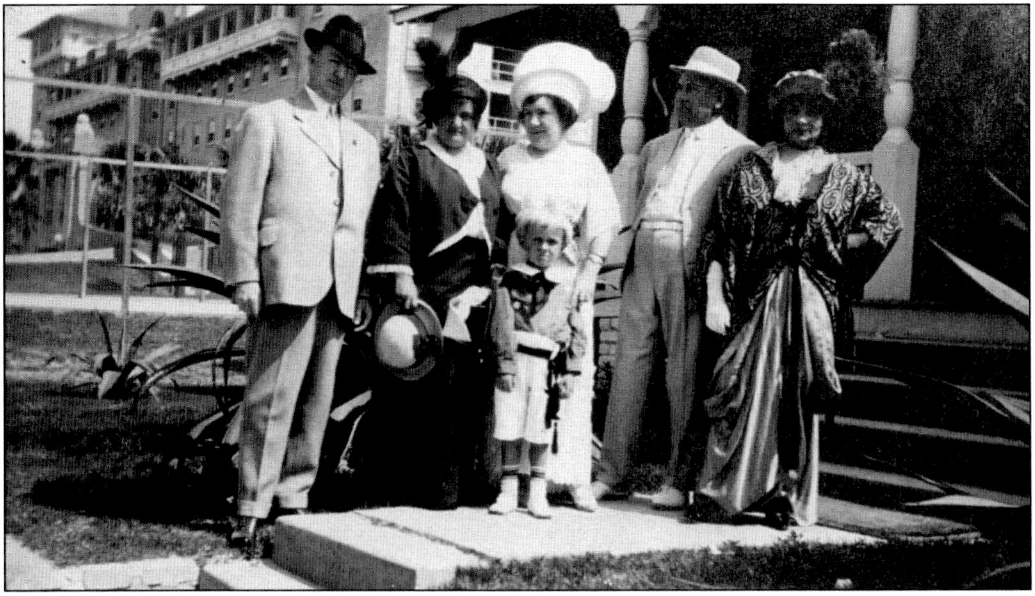

Mike Owens and his wife, along with her sister, are pictured here. The other couple is Tom Rowe and his wife.

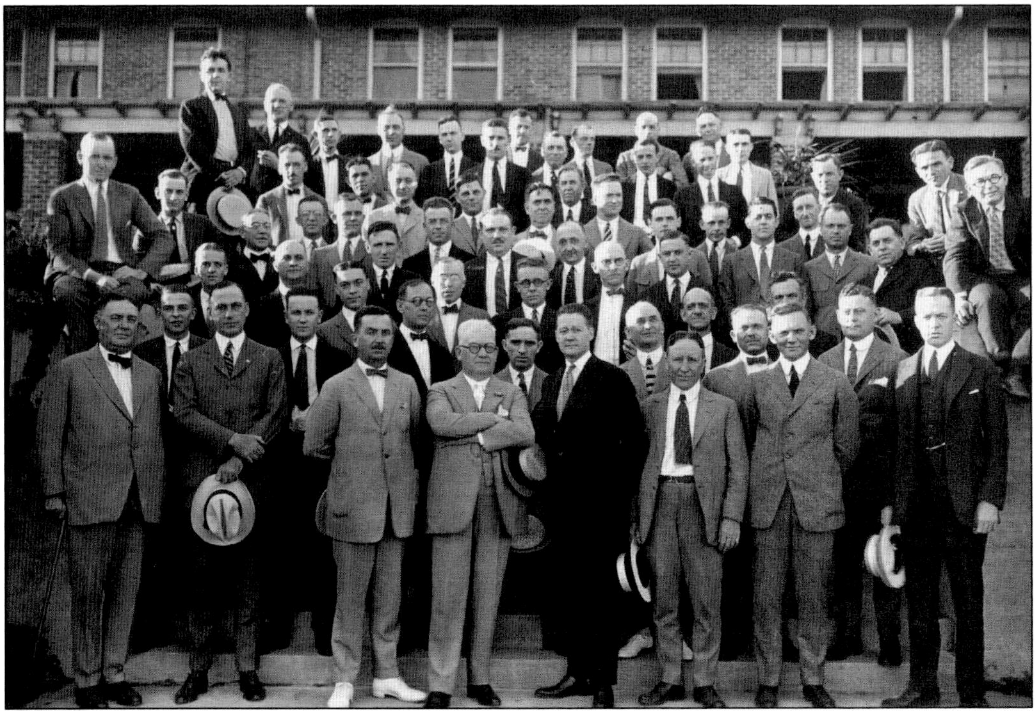

Owens is front row center. Over 95 percent of American bottle production and 80 percent of the world's bottle production was on Owens machines in 1922. Over four million gross of beer, soda, and mineral water bottles, as well as over one million gross of whiskey bottles were produced in America on Owens machines that year. To the left of Owens is William Boshart (future president of Owens Bottle), and to his right is John Biggers (future president of Libbey-Owens-Ford).

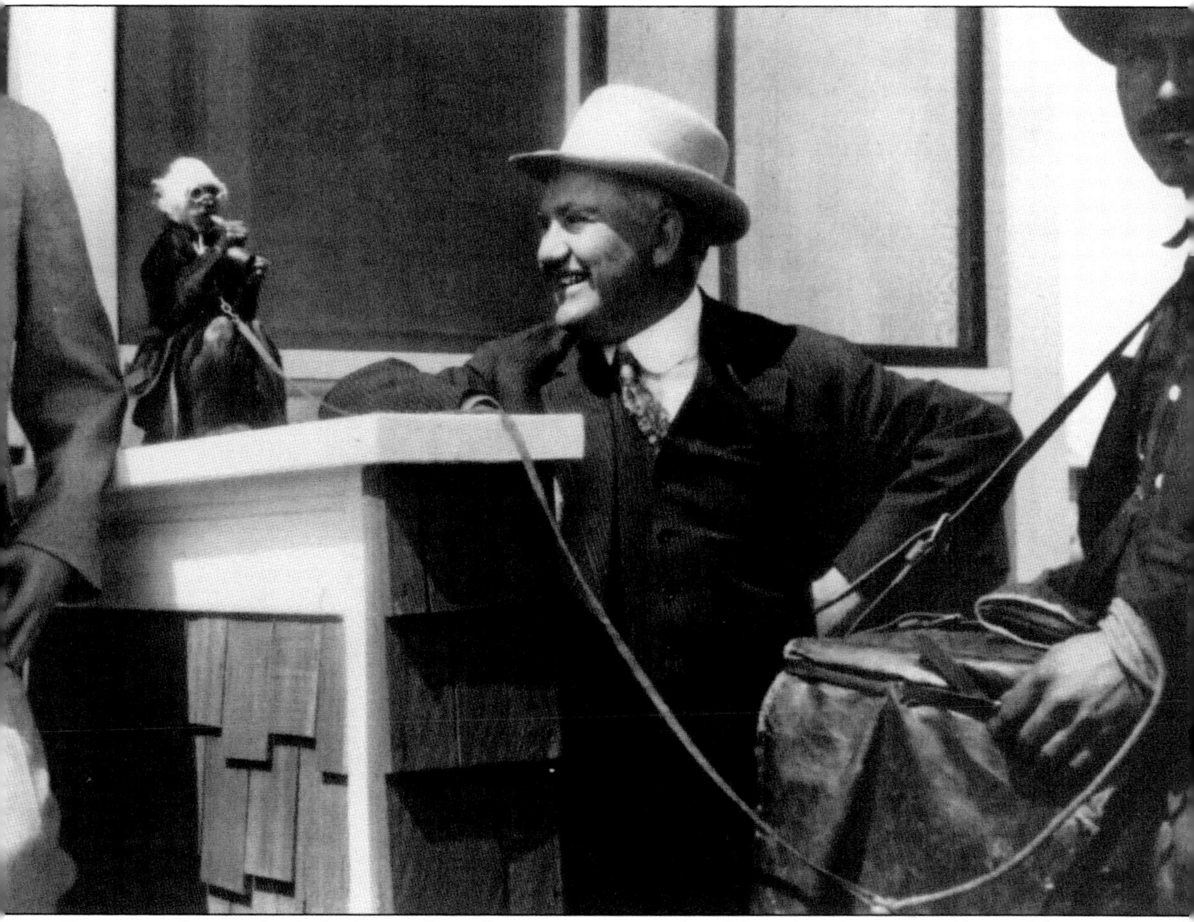

Owens was known both for his smile and temper. He was a loyal friend and a personable neighbor.

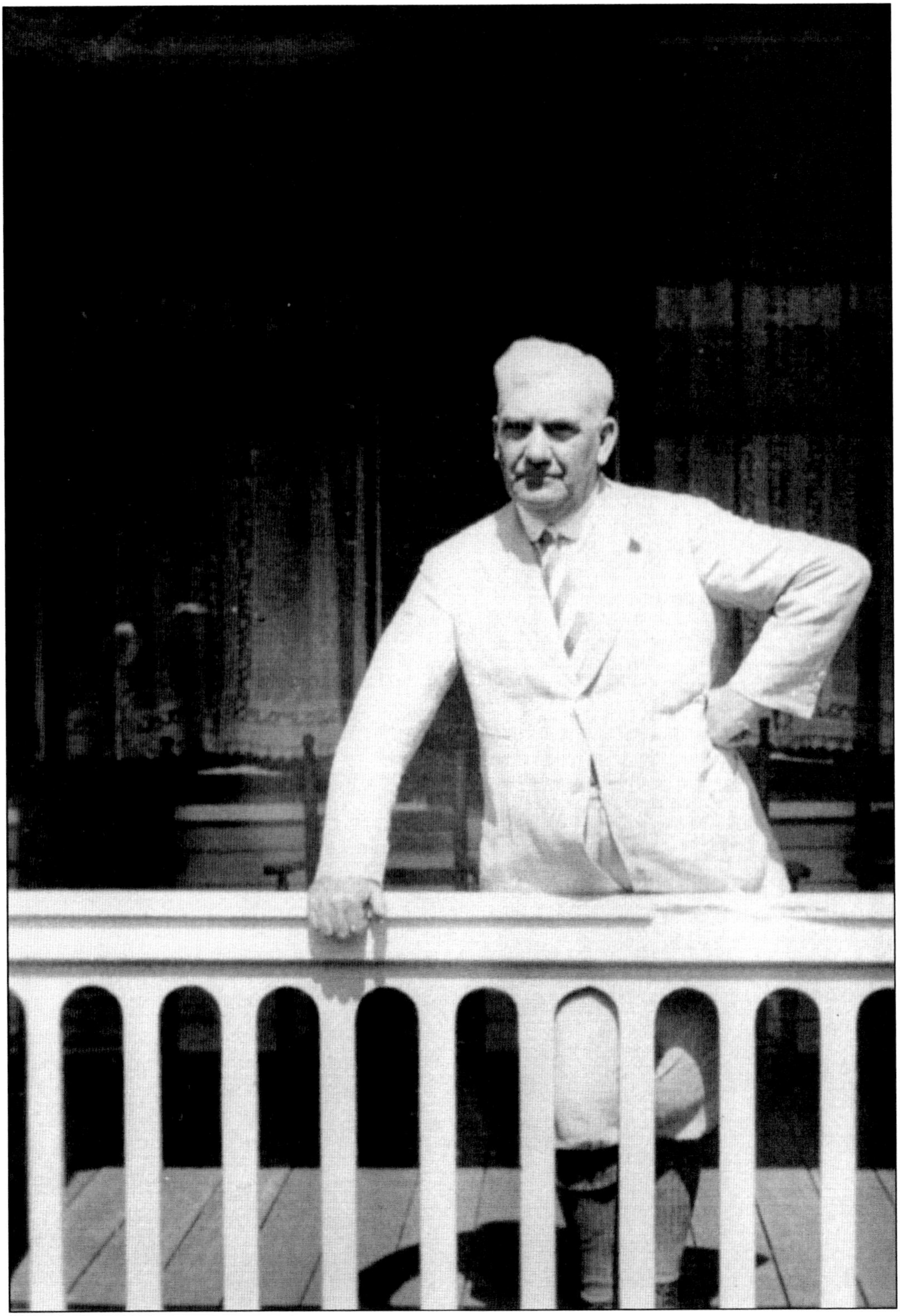

Mike Owens was diagnosed with heart problems in 1917.

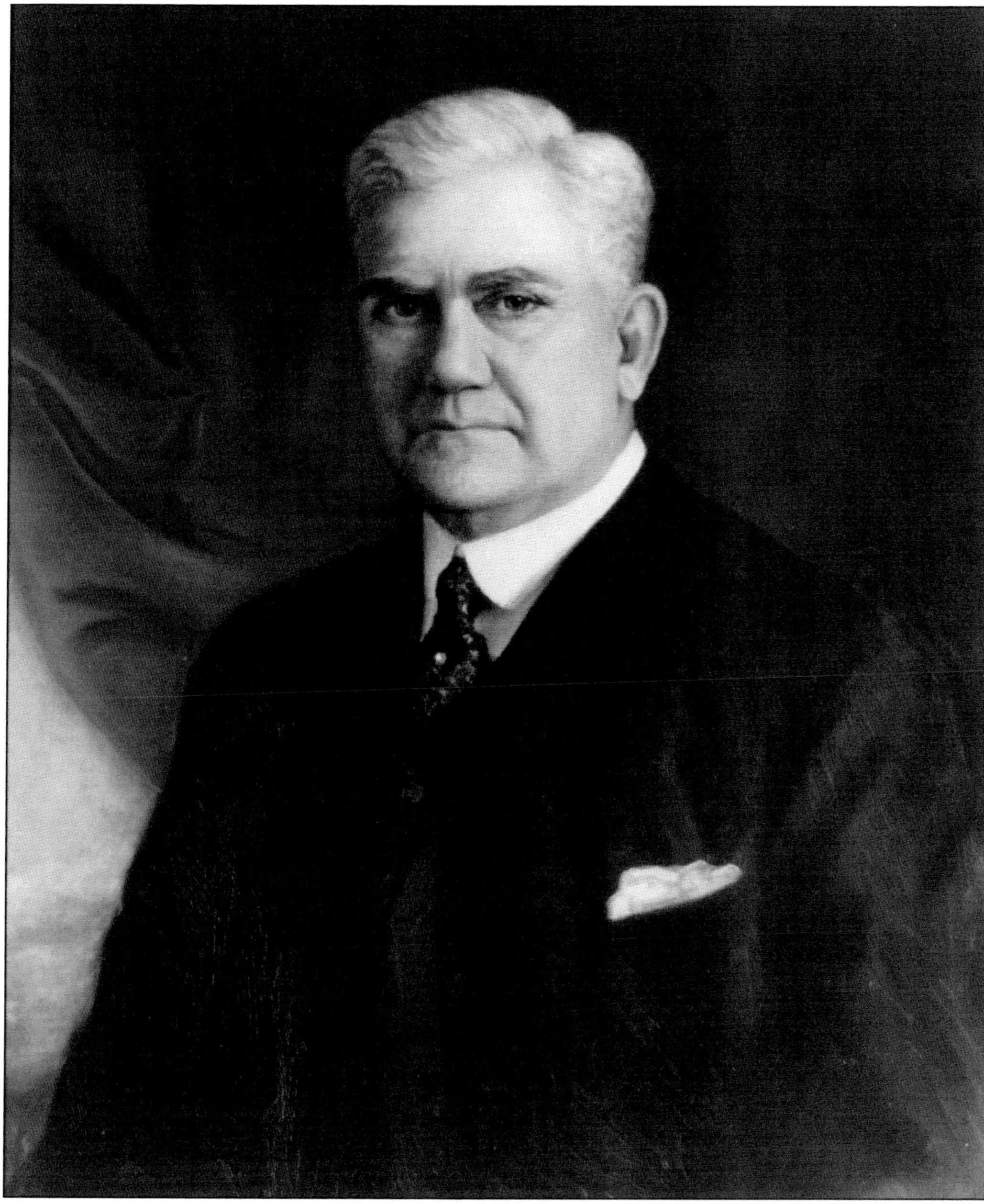

By 1919, Owens was still working but was forced to slow down by his doctors.

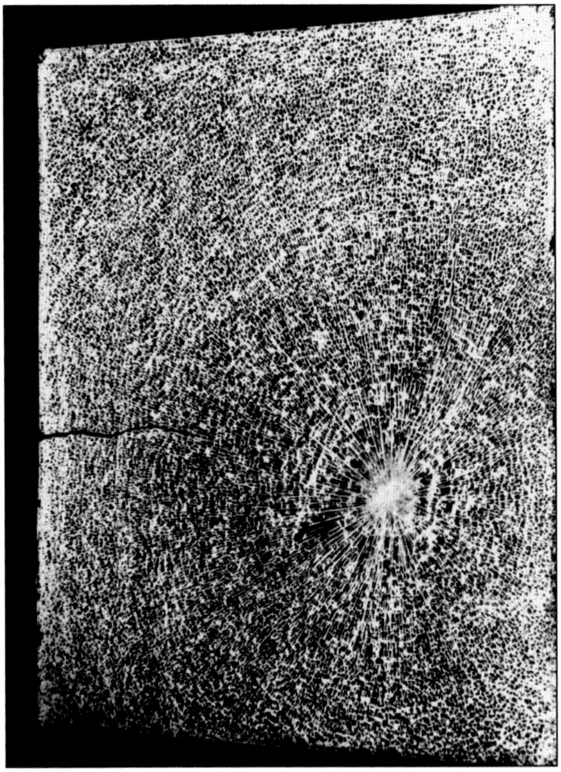

Safety glass for automobiles was one of the projects Mike Owens was working on when he died.

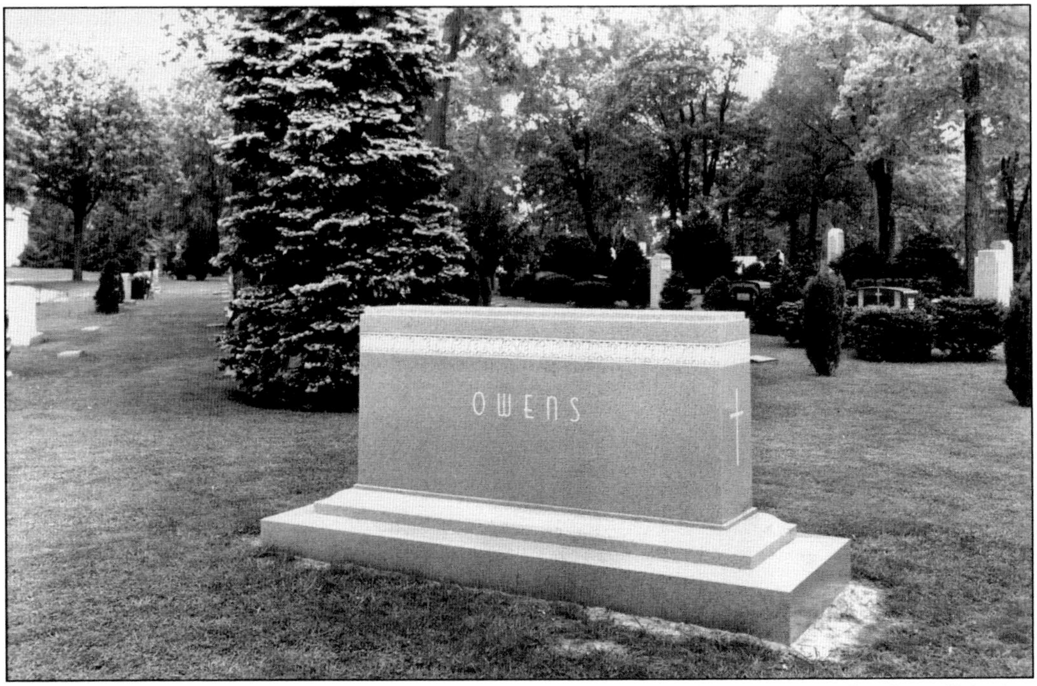

Owens died on December 23, 1923. Public and parochial schools were given the day off for his funeral. The funeral was held at the old Catholic church near today's Rosary Cathedral (for which Owens contributed generously to the building fund).

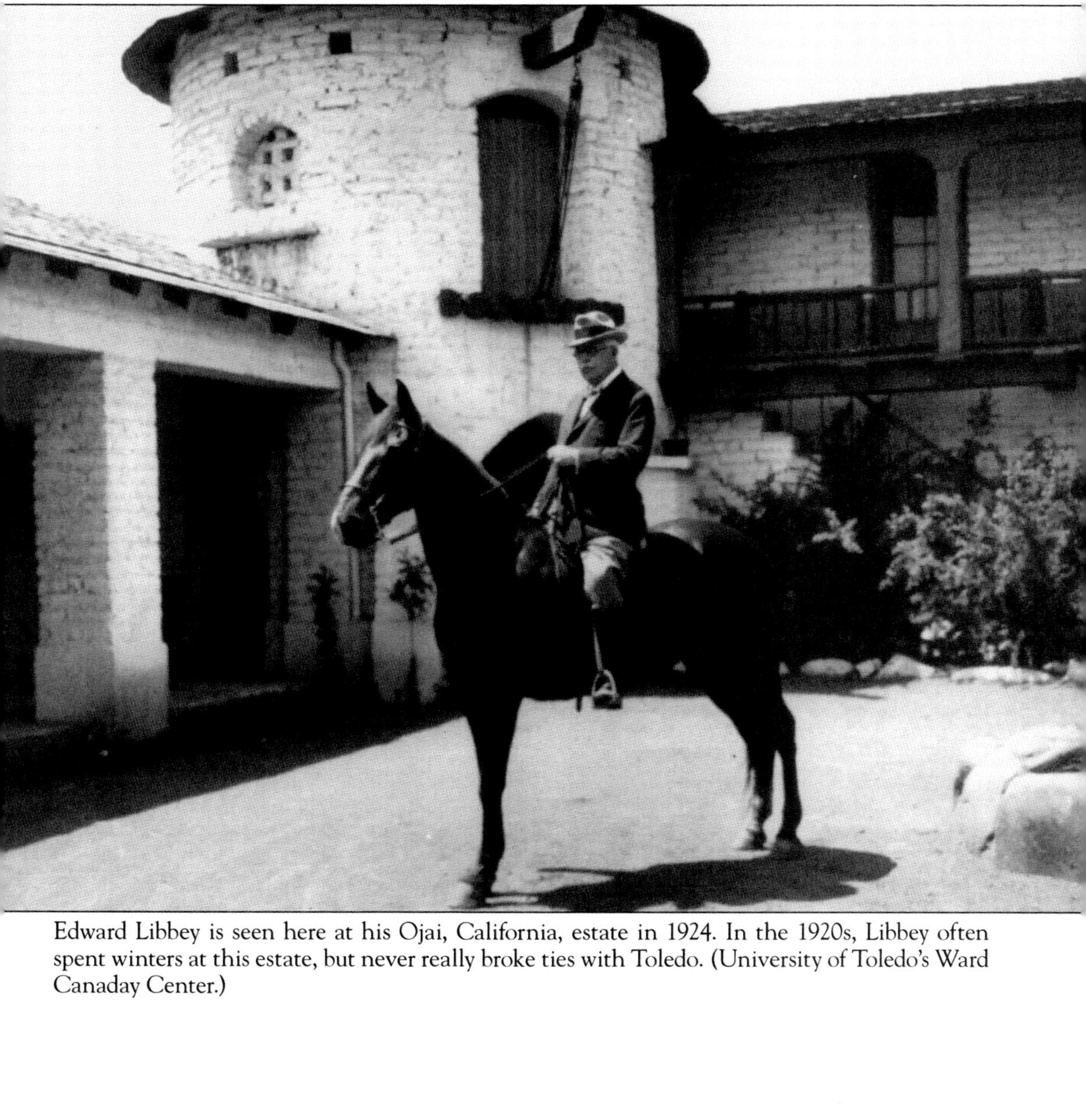

Edward Libbey is seen here at his Ojai, California, estate in 1924. In the 1920s, Libbey often spent winters at this estate, but never really broke ties with Toledo. (University of Toledo's Ward Canaday Center.)

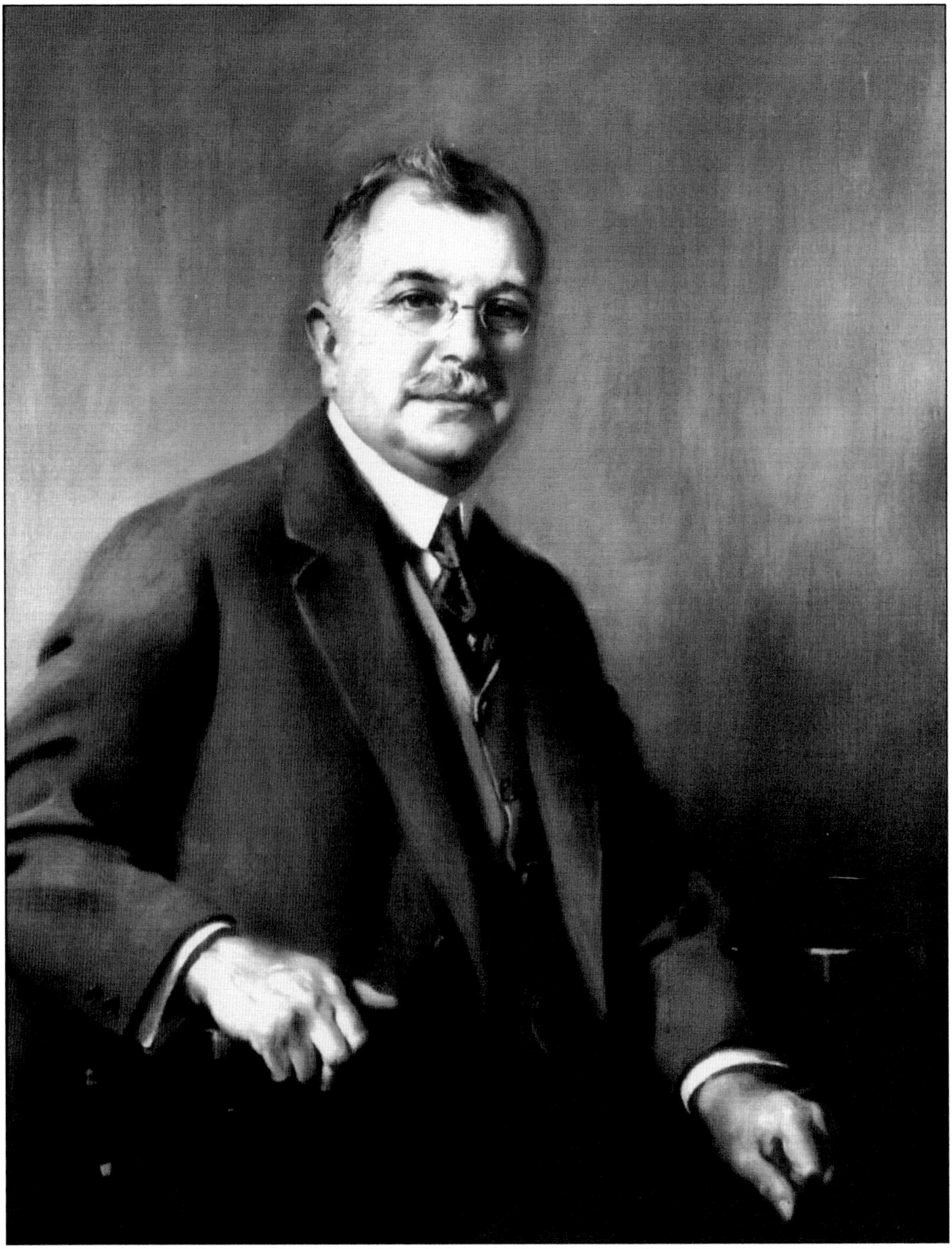

Edward Libbey died of pneumonia after a long trip to Egypt. After Florence Libbey's bequests, most of Libbey's millions went to the Toledo Museum of Art. (Courtesy of the University of Toledo's Ward Canaday Center.)

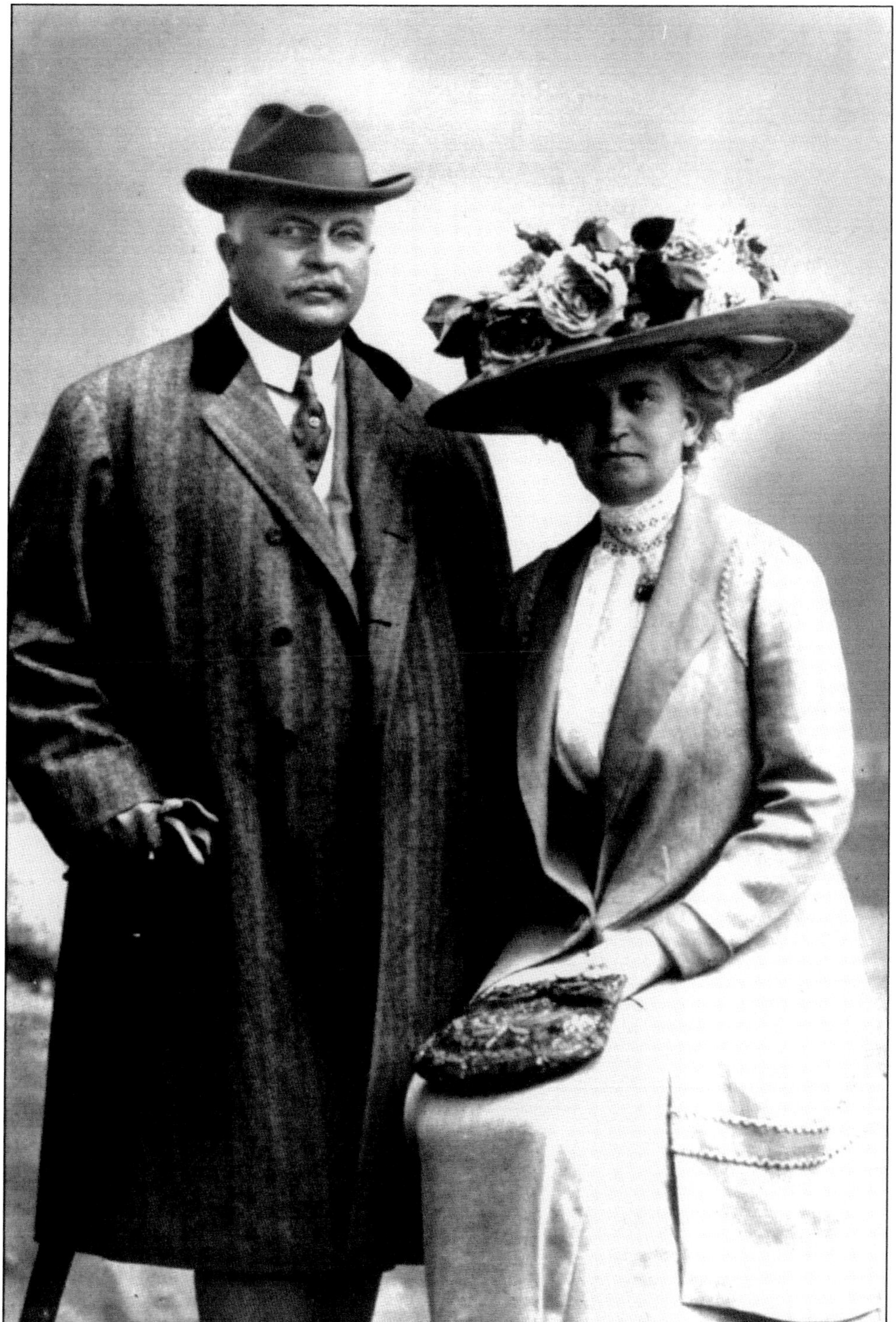

After Edward's death, Florence continued leading the expansion of the Toledo Museum. The Libbeys had no children, and most of their fortunes were in the museum. (Courtesy of the Toledo Museum of Art.)

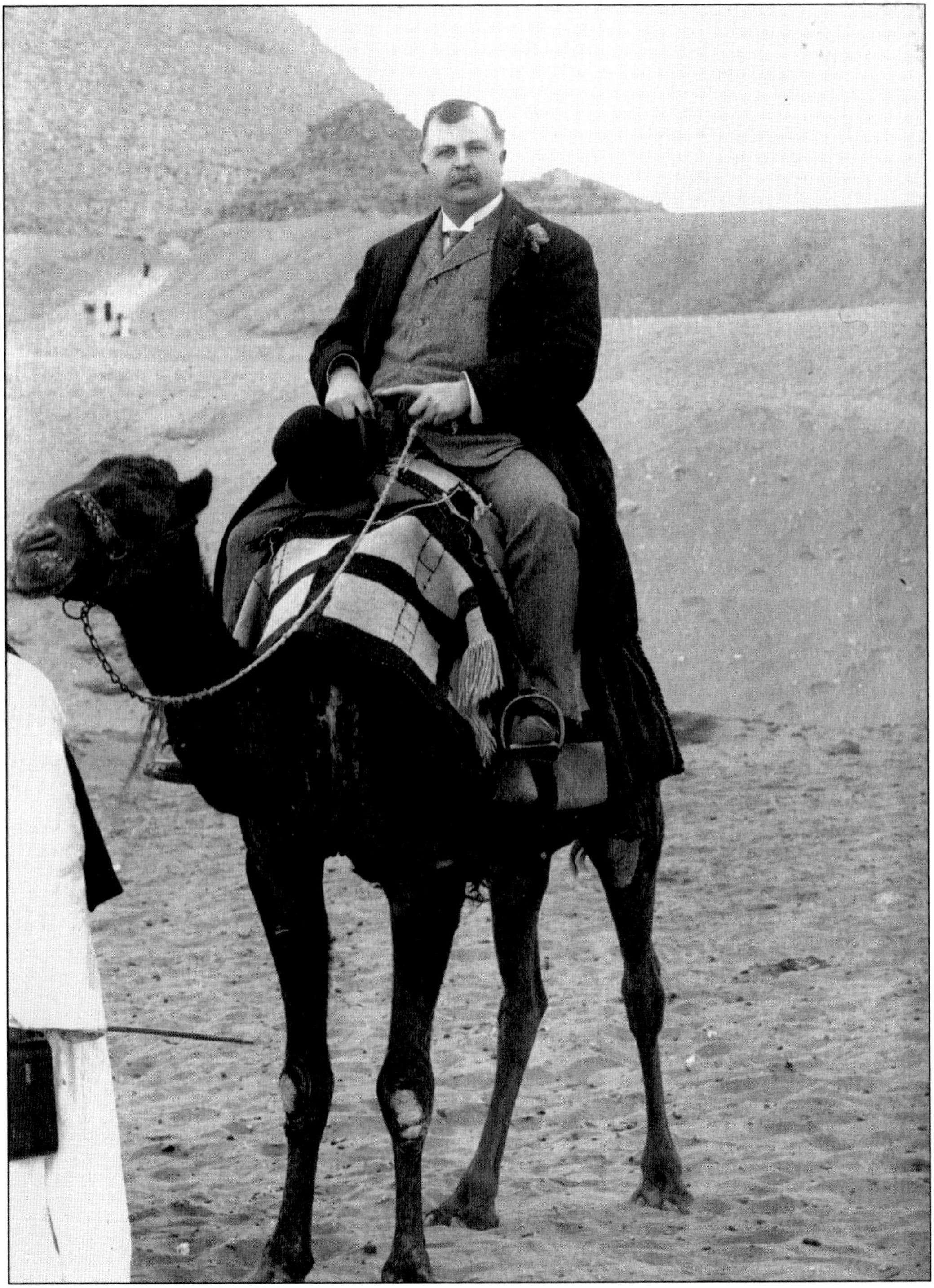

Libbey took many art-collecting trips to Egypt and Europe. His collection became the base of the Toledo Museum of Art. (Courtesy of the Toledo Museum of Art.)

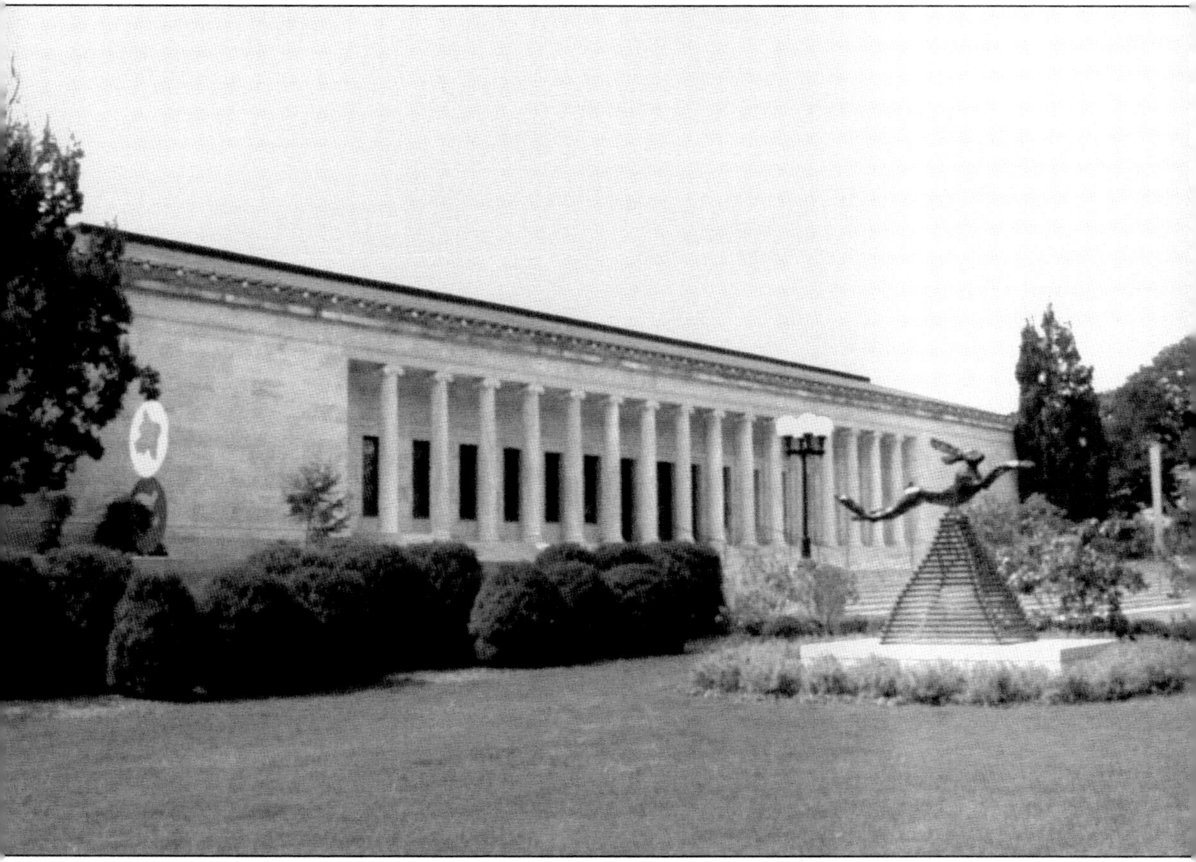

Libbey started his art museum in 1901 using rented rooms in downtown Toledo. In 1903, the museum moved to an old mansion on Thirteenth and Madison Avenues. The central core of the present museum above was built in 1912.

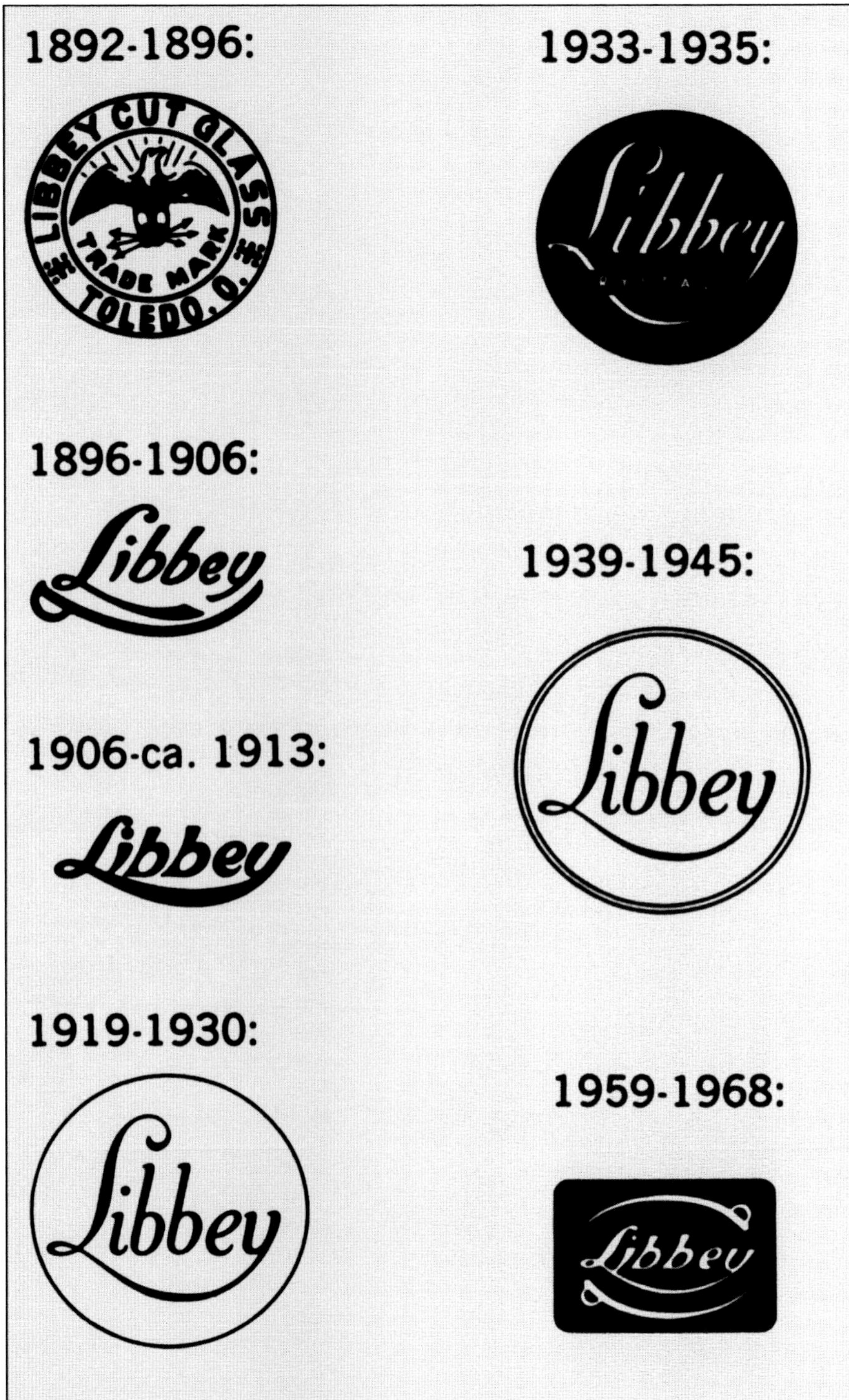

The evolution of the Libbey trademarks is seen here. The sword is a symbol of Toledo, Spain, known for its excellence in sword making. Libbey envisioned the same fame in glass for Toledo, Ohio.

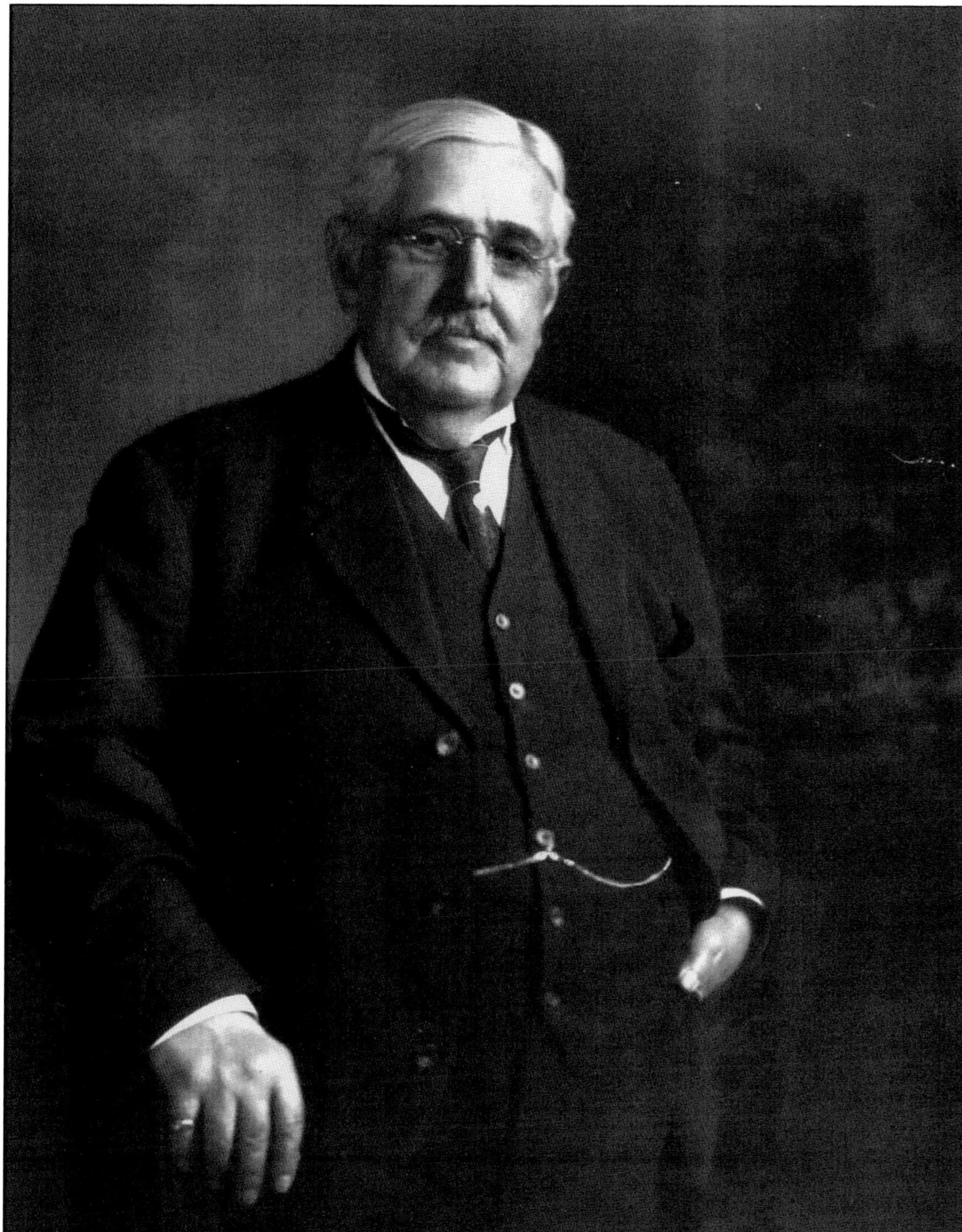

Here is William S. Walbridge, the third of the northwest Ohio glass titans with Libbey and Owens. (Courtesy of the Rossford Public Library.)

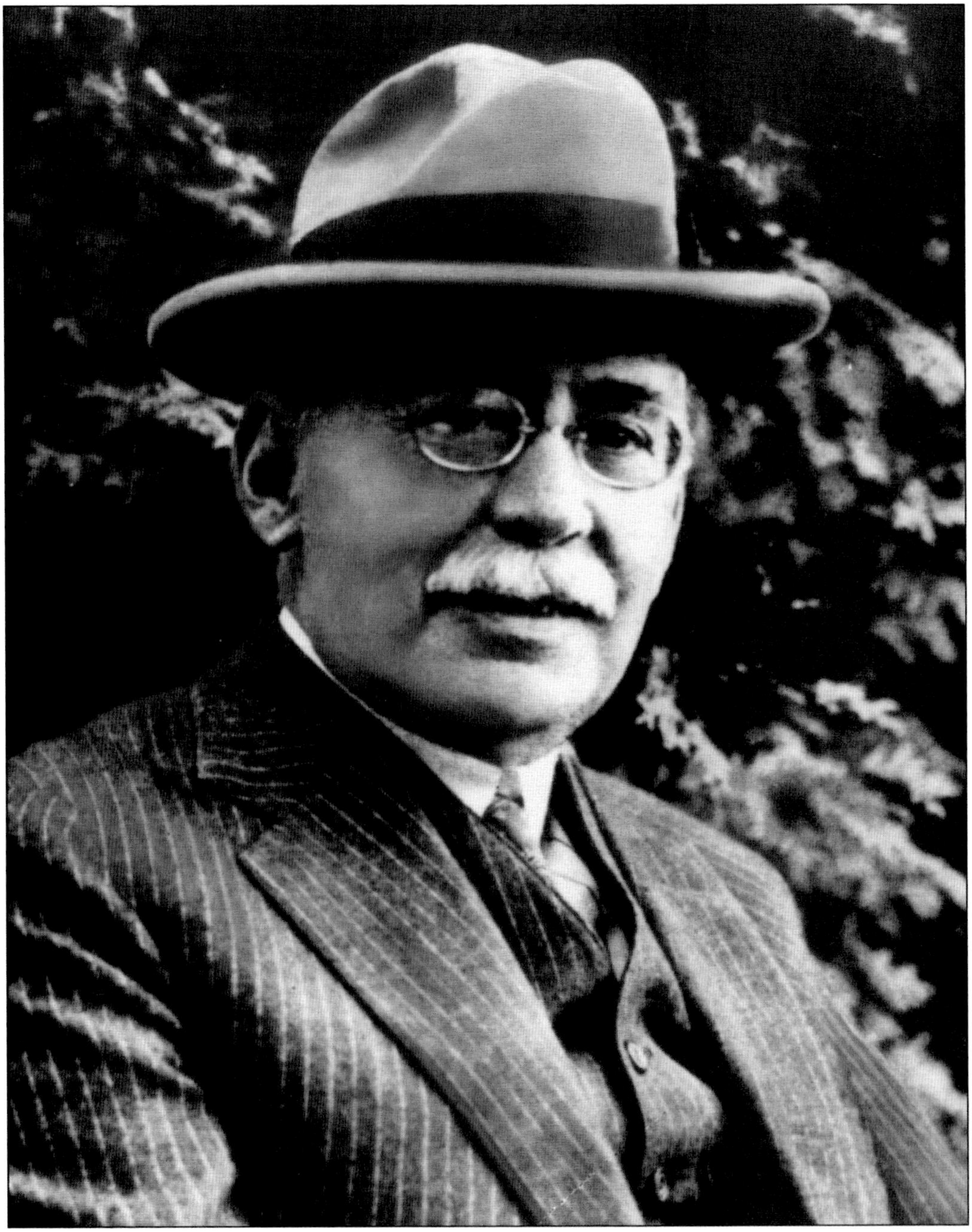

William S. Walbridge was Edward Libbey's brother-in-law and was brought in as treasurer of Libbey Glass, then he became treasurer of Toledo Glass and Owens Bottle Machine Company. Wallbridge would become the world's best snow glass bottle collector and historian. (Courtesy of the University of Toledo's Ward Canaday Center.)

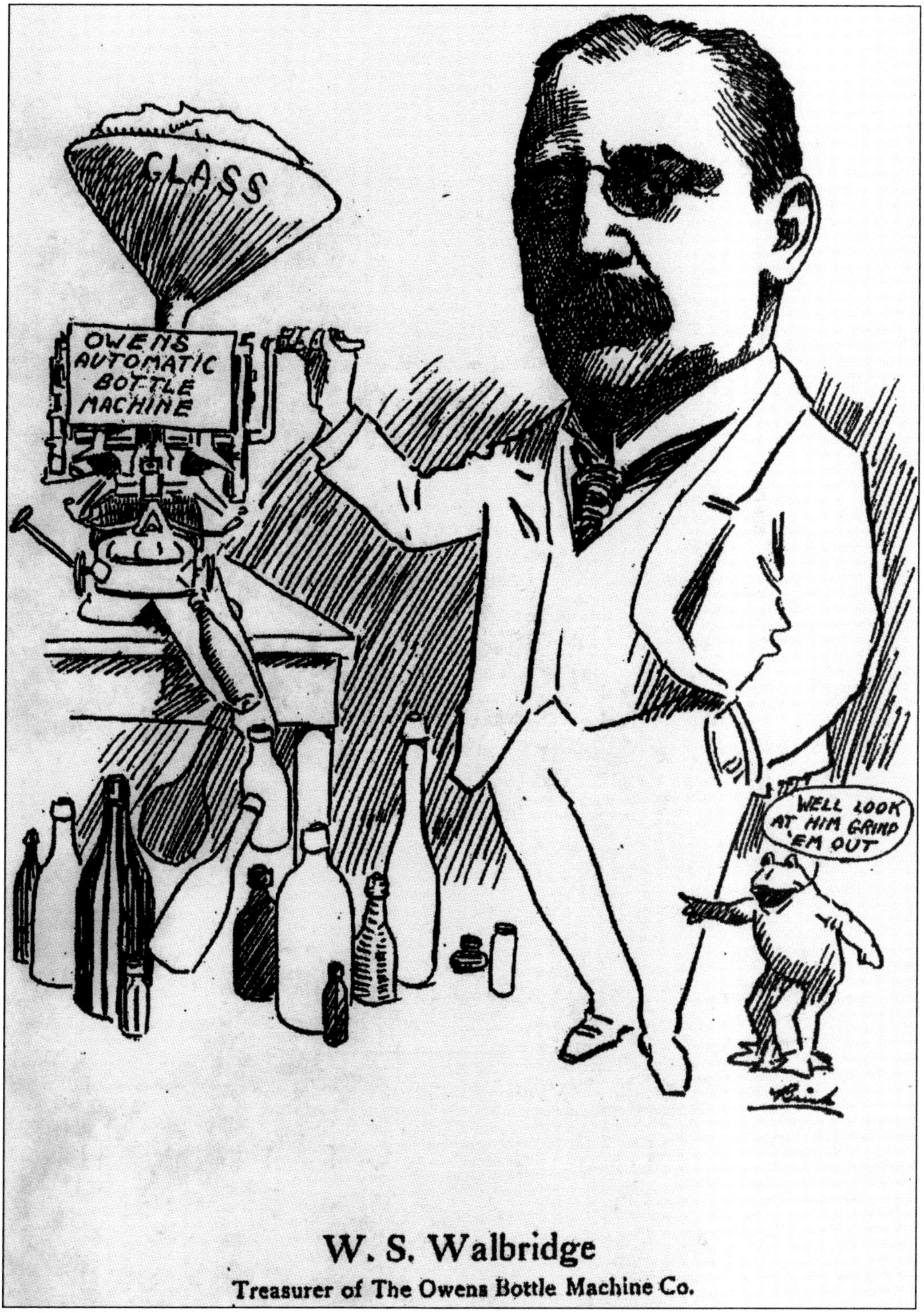

W. S. Walbridge
Treasurer of The Owens Bottle Machine Co.

Walbridge led the sales effort of the machine to the beer industry and did more than anyone to make the Owens machine an international success.

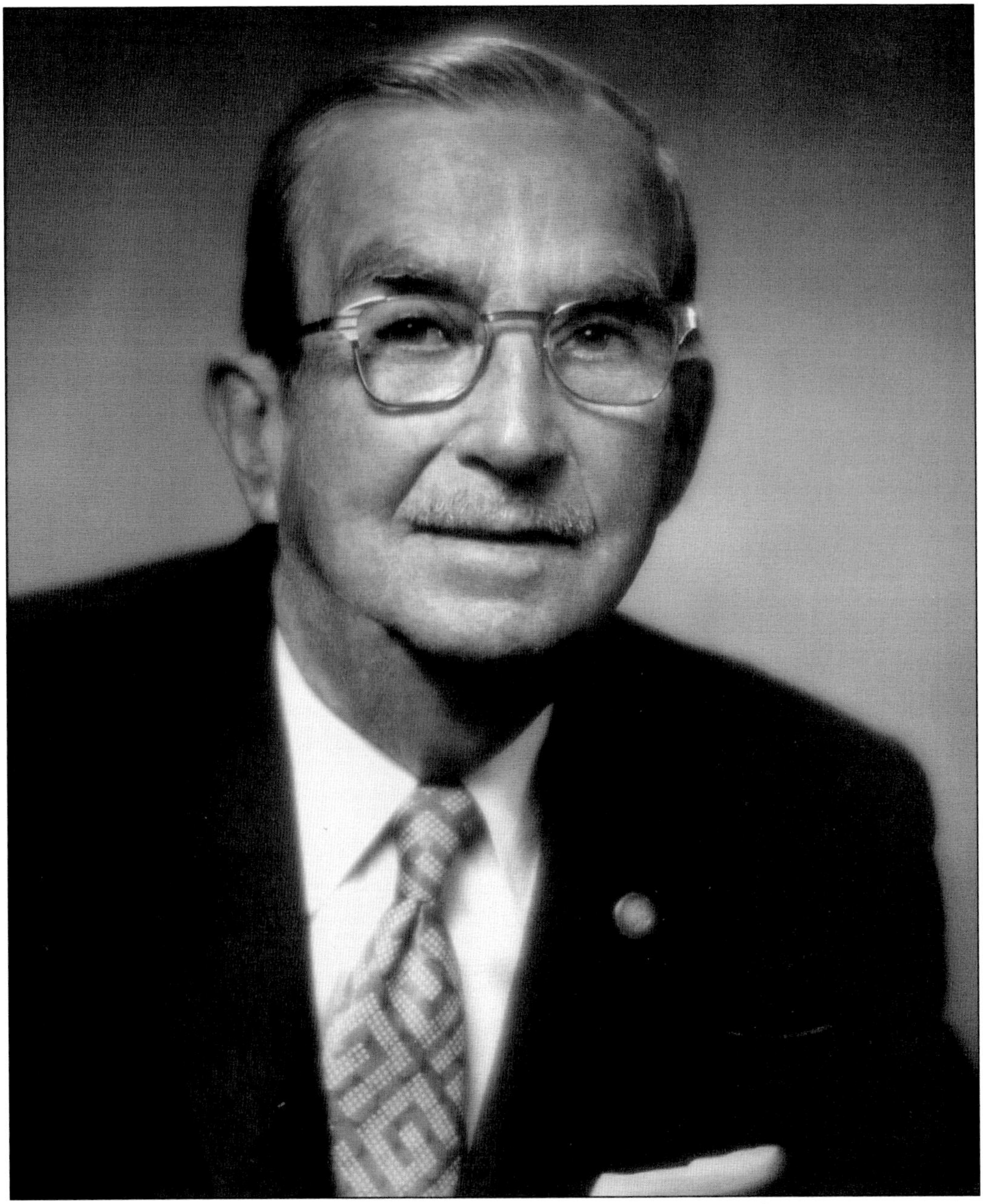

William Boshart joined Toledo Glass in 1906 and became plant manager of European operations of Owens Machine Company in 1913. After Edward Libbey's death in 1925, Boshart became president of Owens Bottle Company, and he formed the merger with Illinois Glass, becoming president of Owens-Illinois in 1929.

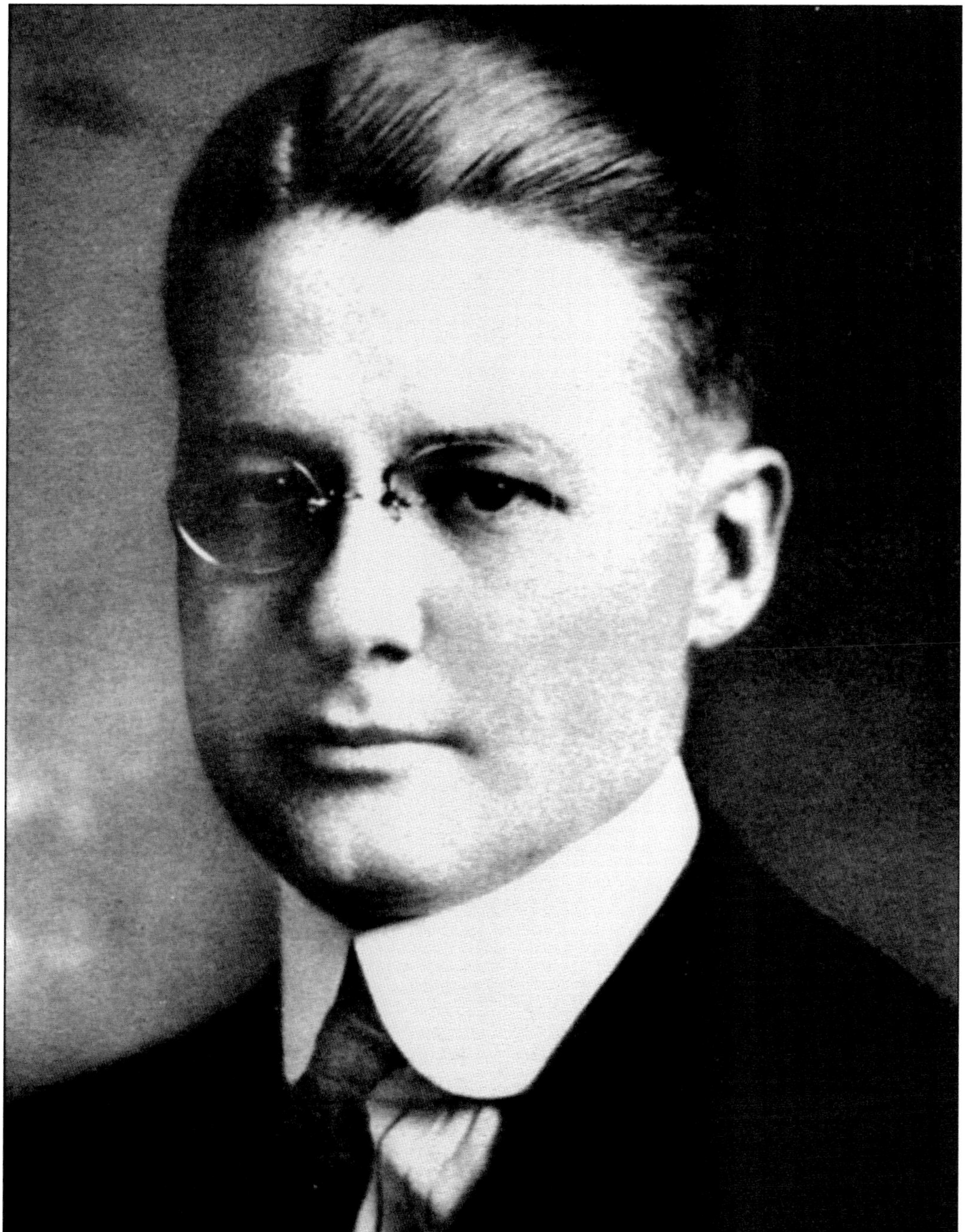

John Biggers graduated from the University of Michigan and started his career at Owens Bottle in 1914 working as an assistant to Michael Owens. He rose to vice president after Owens's death. Later he was one of the directors of Libbey-Owens Sheet Company. (Courtesy of the University of Toledo's Ward Canaday Center.)

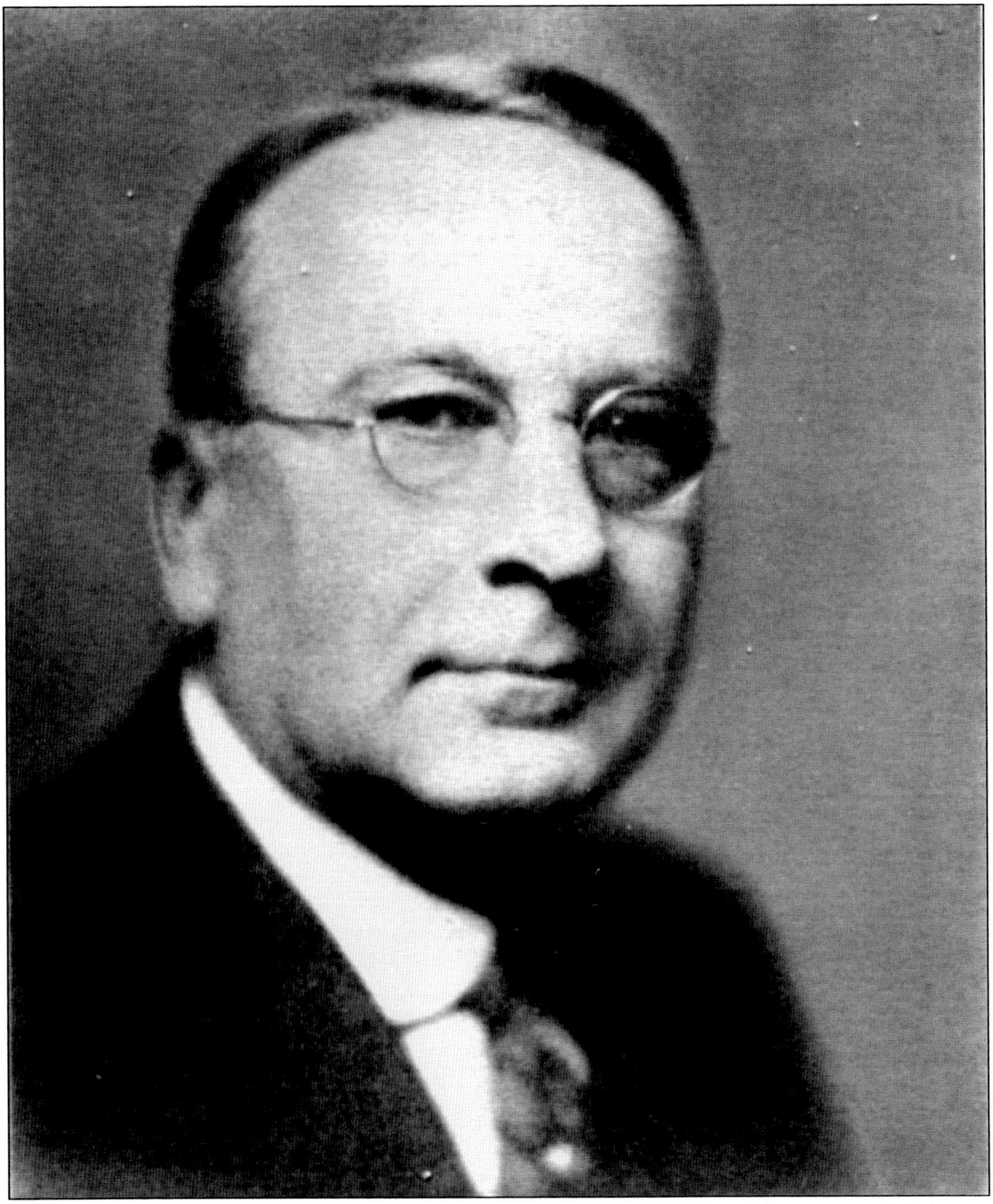

James Blair was the son of Toledoean industrialist Alfred Blair. He was a University of Michigan dentist. Clarence Brown brought him into Owens Bottle management in 1916. He rose to vice president and became a major stockholder. After Edward Libbey's death, Blair took Libbey's throne of glass. In 1930, he engineered the merger and became president of Libbey-Owens-Ford.

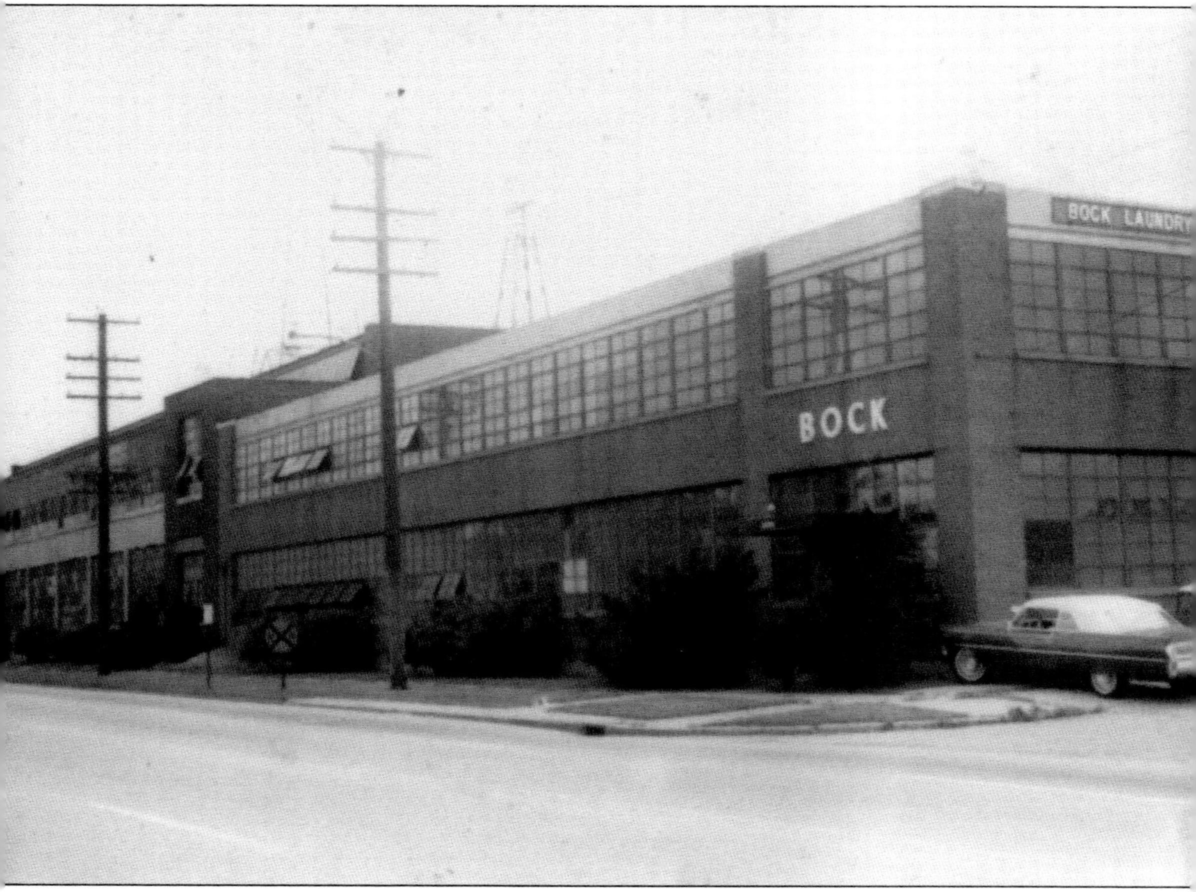

Emil Bock started for Michael Owens at the start of the bottle machine project but resigned in 1910. Bock was a brilliant engineer and industrialist who founded several local industries and machine shops.

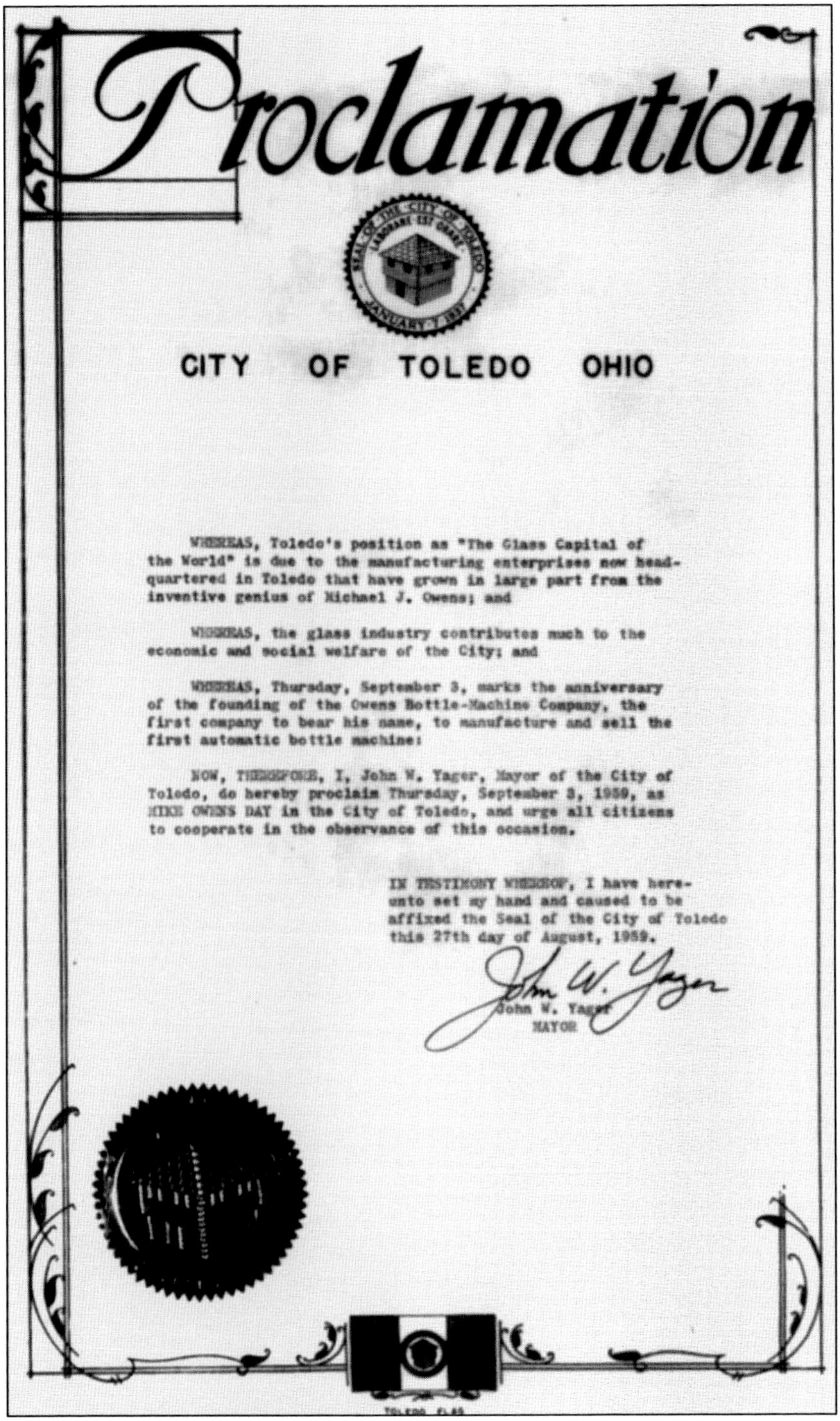
Proclamation

SEAL OF THE CITY OF TOLEDO
LABORARE EST ORARE
JANUARY 7, 1837

CITY OF TOLEDO OHIO

WHEREAS, Toledo's position as "The Glass Capital of the World" is due to the manufacturing enterprises now headquartered in Toledo that have grown in large part from the inventive genius of Michael J. Owens; and

WHEREAS, the glass industry contributes much to the economic and social welfare of the City; and

WHEREAS, Thursday, September 3, marks the anniversary of the founding of the Owens Bottle-Machine Company, the first company to bear his name, to manufacture and sell the first automatic bottle machine;

NOW, THEREFORE, I, John W. Yager, Mayor of the City of Toledo, do hereby proclaim Thursday, September 3, 1959, as MIKE OWENS DAY in the City of Toledo, and urge all citizens to cooperate in the observance of this occasion.

IN TESTIMONY WHEREOF, I have hereunto set my hand and caused to be affixed the Seal of the City of Toledo this 27th day of August, 1959.

John W. Yager
John W. Yager
MAYOR

TOLEDO FLAG

On September 3, a total of 34 Owens-Illinois Glass plants in 33 cities stopped production for a short ceremony to honor Michael Owens. A ceremony was held in Toledo with remarks by John Biggers, chairman of Libbey-Owens-Ford; J. Levis, chairman of Owens-Illinois; and Harold Boeschenstein, chairman of Owens-Corning Fiberglass.

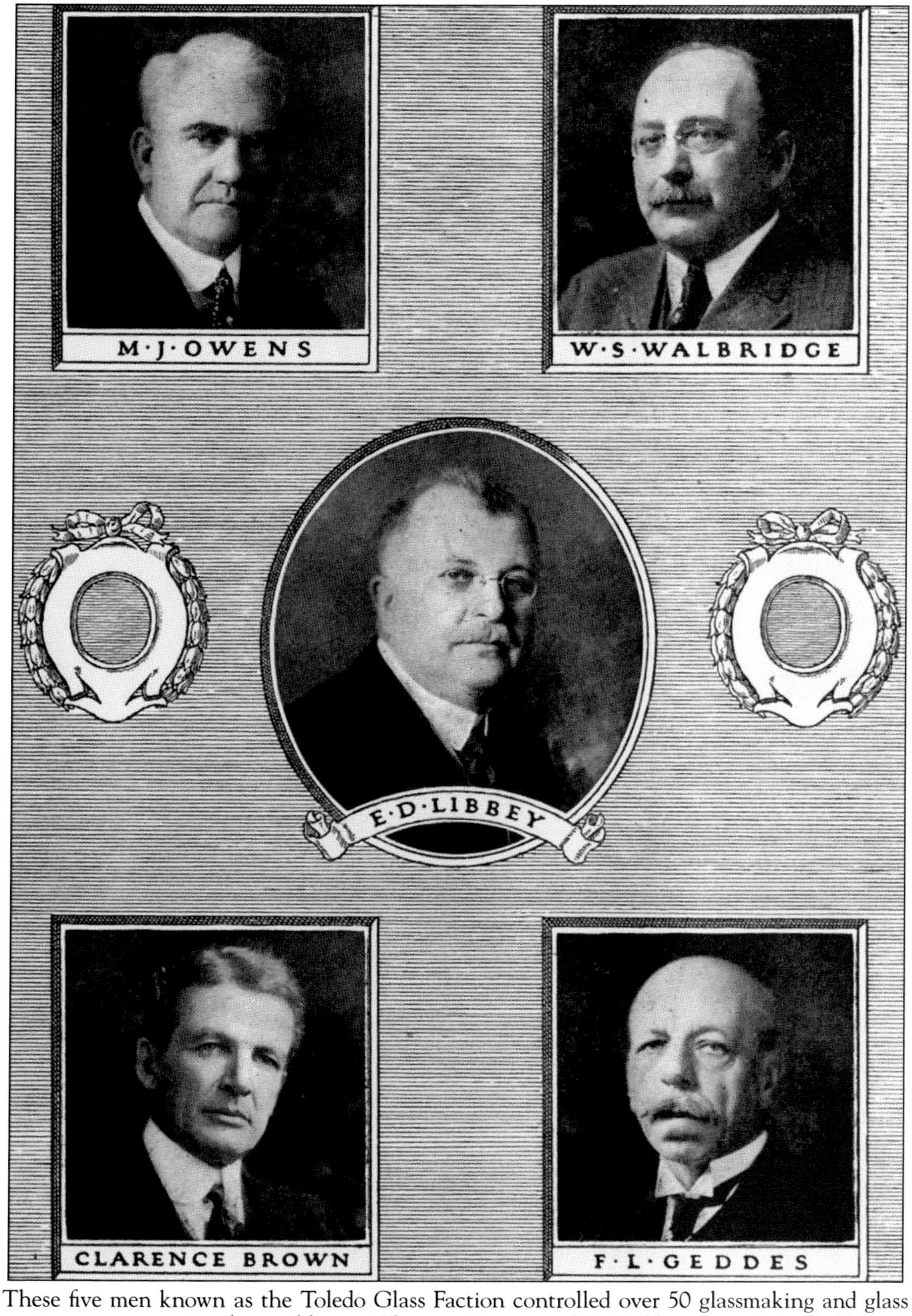

These five men known as the Toledo Glass Faction controlled over 50 glassmaking and glass equipment companies the world over. Clarence Brown and F. Geddes were originally Edward Libbey's lawyers when he came to Toledo.

Discover Thousands of Local History Books Featuring Millions of Vintage Images

Arcadia Publishing, the leading local history publisher in the United States, is committed to making history accessible and meaningful through publishing books that celebrate and preserve the heritage of America's people and places.

Find more books like this at
www.arcadiapublishing.com

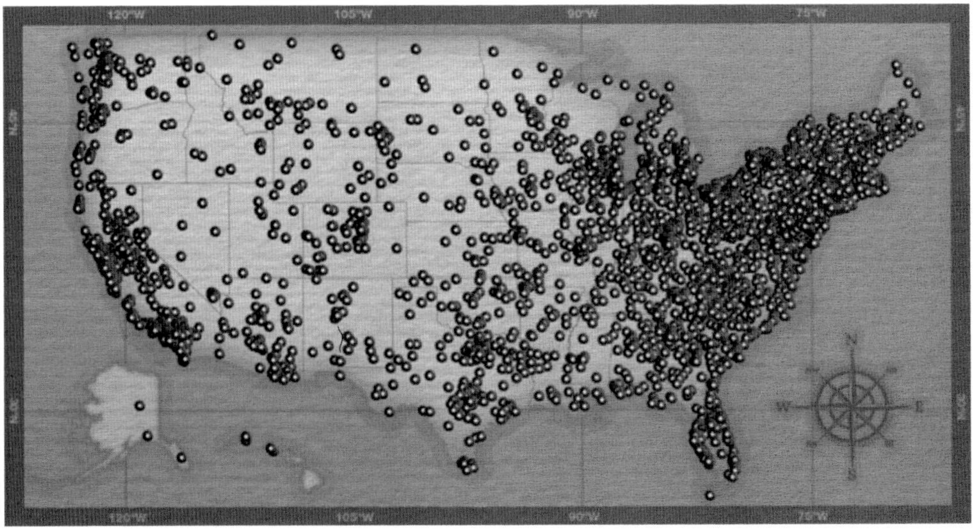

Search for your hometown history, your old stomping grounds, and even your favorite sports team.

Consistent with our mission to preserve history on a local level, this book was printed in South Carolina on American-made paper and manufactured entirely in the United States. Products carrying the accredited Forest Stewardship Council (FSC) label are printed on 100 percent FSC-certified paper.

MADE IN THE USA